An introduction to ethology

An introduction to ethology

P. J. B. SLATER

Kennedy Professor of Natural History, University of St Andrews

The right of the
University of Cambridge
to print and sell
all manner of books
was granted by
Henry VIII in 1534.
The University has printed
and published continuously
since 1584.

CAMBRIDGE UNIVERSITY PRESS

Cambridge

London   New York   New Rochelle

Melbourne   Sydney

Published by the Press Syndicate of the University of Cambridge
The Pitt Building, Trumpington Street, Cambridge CB2 1RP
32 East 57th Street, New York, NY 10022, USA
10 Stamford Road, Oakleigh, Melbourne 3166, Australia

© Cambridge University Press 1985

First published 1985

Printed in Great Britain by the University Press, Cambridge

British Library Cataloguing in Publication Data
Slater, P. J. B.
An introduction to ethology.
1. Animals, Habits and behavior of
I. Title
591.51   QL751

Library of Congress Cataloguing in Publication Data
Slater, Peter J. B.
An introduction to ethology
Bibliography: p.
Includes index.
1. Animal behavior.   I. Title.
QL751.S62   1986   591.51   85-7748

ISBN 0 521 30266 8 hard covers
ISBN 0 521 31605 7 paperback

CONTENTS

Preface   vii

1 **What is ethology?**   1
1.1 A case history   4
1.2 Development and causation   7
1.3 Evolution and function   11
1.4 Book plan   15

2 **Patterns of movement**   16
2.1 Reflexes   16
2.2 Fixed action patterns   20
2.3 Central and peripheral control   25
2.4 Orientation of movements   28

3 **Sensory systems**   34
3.1 Releasers and sign stimuli   37
3.2 What is an egg to a gull?   39
3.3 What the frog's eye tells the frog's brain   41
3.4 Central and peripheral filters   47

4 **Motivation**   51
4.1 A model of motivation   52
4.2 Deciding what to do   65
4.3 Displacement activities   72
4.4 Conclusion   74

5 **Development**   76
5.1 Psychologists and learning   76
5.2 Ethologists and instinct   79
5.3 A false dichotomy   83
5.4 Some case histories   88
5.5 Conclusion   98

6 **Evolution**   100
6.1 Behaviour genetics   100
6.2 Comparative evidence   106
6.3 The origin of displays   112

7 Function 119
7.1 The concept of function 119
7.2 Experiments and observations 122
7.3 The sociobiological revolution 134
7.4 Strategies of reproduction 138

8 Communication 144
8.1 Crypsis and communication 144
8.2 Information and manipulation 150
8.3 Messages and their meanings 154
8.4 The form of signals 157
8.5 Animal communication as language 161

9 Social organisation 165
9.1 Why live in groups? 165
9.2 Kinship 172
9.3 Cooperation 176
9.4 Dominance 178
9.5 Synchrony 181
9.6 Culture 183

Selected reading 187
List of species names 189
Index 191

PREFACE

This book aims to introduce students at advanced school and early university level to the study of animal behaviour. This is an exciting area of research about which, thanks to a great many popular books and television programmes, everyone opening these pages must already know something. However, ideas change rapidly in active fields of science, and ethology is no exception to this. Indeed rather few of the concepts developed by ethologists before 1950 are still found useful by those studying the subject today. Nevertheless, the theory built up at that time was clear and simple and provided a magnificient body of hypotheses on which subsequent work could be based. For this reason, the approach I shall take here is somewhat historical, looking at the ideas of the early ethologists, notably Konrad Lorenz and Niko Tinbergen, and then surveying developments over the past few decades to identify subsequent work which has been influential in leading to what we now think about animal behaviour. I have tried to choose some of the best examples, from amongst both classic studies in the field and the great range of recent work, to illustrate my themes as clearly as possible for those with no previous knowledge of ethology. These have the excitement of approaching this fascinating subject for the first time; if I can share some of the thrill I felt when I did so, this book will have succeeded.

It is a pleasure to thank Jan Parr for the fine illustrations, Sandi Irvine for her careful copy editing and Martin Walters for seeing the book through the presses. Parts of the book have been read by Neil Chalmers, Peter Clifton, Aubrey Manning, Tim Roper and Liz Slater and I am very grateful to them for taking the time to do so. It has gained greatly from their suggestions.

St Andrews   P.J.B.S.
November 1984

1

What is ethology?

The question which forms the title of this chapter is not simply rhetorical, nor is it one to which most dictionaries give a useful answer. Any but the most modern are likely to define ethology as 'the science of character', based on a use of the term dating from the last century. This neglect of its modern meaning is, quite simply, because ethology is a very young branch of science and most reference books have not yet recognised its existence. In its current meaning, ethology is what Niko Tinbergen, one of its most famous experts, referred to as 'the biological study of behaviour'.

Of course biologists have studied behaviour for centuries but, as with so much else in biology, interest in behaviour received its most important boost from the writings of Charles Darwin. Darwin included a chapter on 'Instinct' in *The Origin of Species* and he also wrote a book specifically about behaviour called *The Expression of the Emotions in Man and in Animals*. However, in the half-century after Darwin, there was little work on the behaviour of animals, while zoologists grappled with trying to understand the fundamental principles of systematics, physiology and developmental biology. A few scientists from that time, like Julian Huxley in Britain, Oscar Heinroth in Germany and Charles Otis Whitman in America, stand out for their contribution to behaviour, but they were a small band. It was only in the 1930s that a comprehensive theory of animal behaviour began to emerge through the writings of Konrad Lorenz and, later, of Niko Tinbergen. In those years, ethology can truly be said to have been born. Then, in 1972, it came of age as a science when Lorenz and Tinbergen received the Nobel Prize for physiology. They shared it with Karl von Frisch, who discovered the remarkable dance of the honey-bee, which enables foragers to tell others in their hive the location of good food sources. This prize was recognition indeed that these three men, whom many saw simply as naturalists, had made a fundamental and lasting contribution to science (Figure 1.1).

It is, in fact, excusable to think of ethology as being a branch of natural

Figure 1.1. Ethology's three Nobel prize winners: (*a*) Konrad Lorenz (photograph by H. Kacher), (*b*) Niko Tinbergen (photograph by B. Tschanz) and (*c*) Karl von Frisch (photograph by M. von Frisch).

Figure 1.2. Though many ethologists today study their animals in the controlled environment of the laboratory, the essence of the subject is to understand behaviour as it occurs in nature and, for this reason, much ethological research is still conducted in the wild (photograph © K. J. Stewart).

history, for the diversity of nature has always been a source of interest and wonderment to those who, like ethologists, study the natural behaviour of animals. Observing and describing exactly what animals do is a fascinating study in its own right, and it is also an essential prelude to a more scientific analysis of their behaviour. Thus many ethologists spend long hours patiently watching their animals (Figure 1.2) and this can, in itself, be quite revealing even if it is not extended to forming hypotheses and carrying out experiments.

As a result of this preliminary period of thorough and careful description, ethologists are able to make an inventory, or ethogram, of the behaviour patterns possessed by each species that they study. To the casual observer it might look as though different species of birds or of fishes behave in much the same way. It might also seem that the behaviour of each animal is a highly varied business, not easy to split up into particular types. Fortunately, for most animal species, these impressions are not totally true. Each species tends to have an array of stereotyped behaviour patterns, some of which may be shared with related species but others of which are unique to itself.

Describing them and recognising them each time they appear is not as difficult as it might at first appear.

1.1 A case history

Let us illustrate this point with an old favourite of behavioural studies, the three-spined stickleback, an especially easy species to study as it behaves more or less normally in an aquarium tank. Male sticklebacks come into breeding condition in the spring when daylength increases and the streams in which they live become warmer. They change colour, becoming bright red on the underside and iridescent blue on the face, and their behaviour also alters. They gather weed and collect it at a particular spot on the bottom of their pond, gluing it together to form a nest, with a special movement that extrudes a sticky secretion from their cloaca. If another male approaches, the territory owner will chase him away or, if he persists, threaten him by adopting a head-down posture which shows off the red belly in all its brilliance (Figure 1.3*a*). Signals such as this are known as displays. If a ripe female stickleback appears, her belly swollen with eggs, our territory owner behaves quite differently, showing another display known as the zig-zag dance (Figure 1.3*b*). He darts alternately towards and away from the female in a very striking manner and, if she follows him, he draws her slowly towards his nest. Once there, she may creep through the nest and spawn, and he will then follow, fertilising the eggs she has produced as he does so. Her part is then over; indeed he is likely to chase her away, for care of the eggs and of the young after they hatch is, in sticklebacks, carried out by the male alone. When he has eggs he stays close by the nest and repairs any damage that it may suffer, as well as showing fanning, a movement which serves to drive a stream of water over the eggs and so keep them supplied with oxygen (Figure 1.3*c*).

   This description allows the identification of certain behaviour patterns which are common to all male three-spined sticklebacks in breeding condition: 'gluing', 'the head-down threat posture', 'the zig-zag dance', 'creeping through', 'fanning'. All these would appear in an ethogram of this species. But the description also raises a great many questions, and it is here that the scientific aspect of ethology begins. Niko Tinbergen recognised that the sorts of questions one could ask about behaviour fell into four different categories: those about development, causes, functions and evolution. Examples of studies examining these four different categories will be shown in Figs. 1.5–1.8. Interest in development might lead one to ask how a male comes to behave in the way that he does during the course

Figure 1.3. Three of the behaviour patterns shown by a male three-spined stickleback in breeding condition: (*a*) the head-down threat display shown to other males that intrude on his territory; (*b*) the zig-zag dance with which he leads a ripe female to his nest; (*c*) fanning at the nest which he shows mainly after the eggs have been laid.

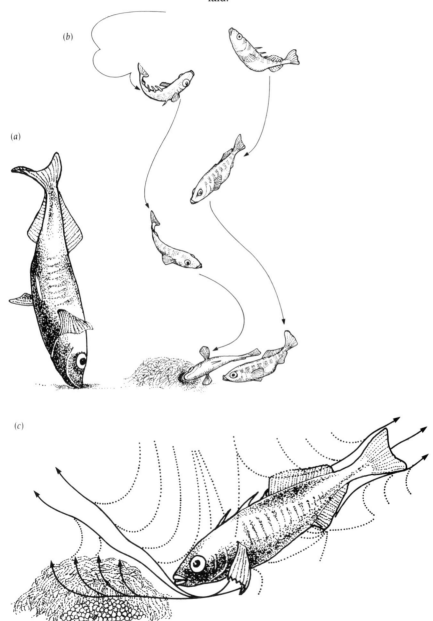

of his life-time. For example, does his skill at nest building improve with practice? Does he court a female the very first time that he sees one or must he learn that this is an appropriate way to behave towards her? On the subject of causes, one can ask about both internal states and external stimuli. What is it that signals the male to come into breeding condition in the spring, and how does this affect his physiology so that he is ready to fight and to court? He behaves differently towards females and towards other males: what difference between them leads him to do so? Functional questions concern the advantages of behaving in a particular way. Why does the male show the zig-zag dance rather than simply swimming to the nest? Does his head-down threat posture actually deter other males from approaching, as it should do if its function is to act as a threat signal? Finally, we can ask about the evolution of behaviour. Comparison between different species of sticklebacks can give us clues about how ancient or recent are particular forms of behaviour. Comparison between displays and other behaviour patterns may suggest what actions formed the originals from which displays have been derived.

This account of the topics with which ethology is concerned makes it sound as though the subject is straightforward, attacking well-defined problems and without any great controversy. Such an impression, however, would be totally wrong. Ethologists have had to contend with gales blowing from various different directions during their short voyage, as well as with awkward questions from some of their own number determined to rock the boat. This has been no bad thing. The early ethologists put forward sweeping general theories based more on careful observation and brilliant intuition than on thorough experimental evidence. These formed a marvellous source of hypotheses for later research, but it was inevitable that this research would lead many of the earlier ideas to be substantially modified or even abandoned altogether. Indeed one ethologist remarked a few years ago that 'ethological theory has diminished at a rate that some find alarming'! Theories might indeed have been disappearing, but it was under a weight of evidence gleaned by enthusiastic ethologists studying a wide range of species in meticulous detail. As a result of this work more detailed knowledge has accumulated and the broad and simple theories have had to be replaced. Ethologists have also developed a reluctance to generalise, as it has become clear that different animal species vary a great deal in their behaviour so that all-embracing theories are not likely to be very helpful. Rats are not just large mice, far less small people!

1.2 Development and causation

The first real storm to hit ethology came in a confrontation with the American school of comparative psychology. The two groups shared an interest in the behaviour of animals, but they approached it from very different viewpoints. The ethologists worked largely in continental Europe and, being zoologists, they had a respect for evolution and were thus interested in a wide variety of species and the different ways in which they behaved. Despite their name, the comparative psychologists were not concerned with such comparisons and tended to study very few species, usually just rats and pigeons, their interest being to look for general laws of behaviour that would hold regardless of the species beings studied, and preferably apply to man as well. Their reputation was for rigorous experimental work in carefully controlled laboratory conditions; most

Figure 1.4. A traditional view of the distinction between ethology and psychology. The psychologist puts his animal in a small box and peers in to see what it is doing, while the ethologist puts himself in the box and looks out at what the animals round about are up to.

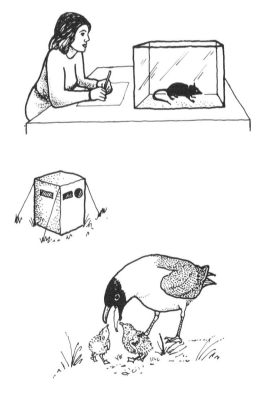

ethologists, on the other hand, simply observed their subjects behaving freely and they did so in the totally uncontrolled conditions of the animal's natural surroundings, those to which selection had adapted it (Figure 1.4). That two such different approaches to very similar topics should lead to confrontation is not surprising. The battle was fought over the subject of behaviour development, a subject on which the views of the two schools were especially starkly contrasted.

The different views of development of the ethologists and the comparative psychologists stemmed, in essence, from the fact that one stressed nature and the other nurture. To most psychologists the learning ability of animals and the flexibility it gives their behaviour is the main interest in studying them, for these aspects may shed light on the equivalent attributes of humans. Hence their stress on nurture. To ethologists, on the other hand, the study of species-typical behaviour was a prime concern: they therefore tended to concentrate on patterns which were highly stereotyped and of similar form throughout a species, and they often referred to such acts as 'innate' or 'instinctive'. The assumption here was that nature was all important and nurture was of little consequence. Indeed, Konrad Lorenz once remarked that the developmental origin of behaviour was a subject of more interest to embryologists than to ethologists.

This controversy was a bitter one, but it was also fruitful, for each side had much to gain from the other, and its resolution brought them closer

Figure 1.5. Studies of development trace how behaviour changes during the life-time of the individual. In mammals such as rhesus monkeys, detailed studies of the relationship between mothers and their infants have revealed subtle changes which take place during the period of lactation. At first the mother is very protective and will not let her baby stray; later she rebuffs the baby's advances and so ncourages its independence. Differences in the relationships between mother–infant pairs may have profound effects on how the behaviour of the infant develops.

together. Psychologists came to recognise that evolution had led animal species to be different from each other and placed constraints on what each could learn. For their part, ethologists came to reject the idea that any behaviour was fixed and inflexible and to realise that the acts they studied, no matter how stereotyped, may have been influenced by learning and by other environmental influences. They also came to appreciate the merits of a carefully controlled experimental approach. So, today, many ethologists work in the laboratory and some of them even use the sorts of equipment developed by psychologists for the study of learning, adapted to shed light on ethological questions. In the battle over nature and nurture, the result has been a compromise: both sides have gained from the realisation that neither nature nor nurture can be ignored in the development of any behaviour pattern. The borderline between ethology and psychology, once hotly contested, has now broken down and those trained in either field may be found working on a variety of topics of interest to both.

Development is just one area of mutual interest to ethologists and psychologists. Another is in the field of causation, the study of those outside influences and internal states that lead animals to behave in the way that they do. The senses of animals keep them informed about changes in the

Figure 1.6. An example of the study of causation. If placed on her own with a nest and eggs a female Barbary dove will not immediately incubate; she needs several days of courtship from a male before she becomes broody. Experiments have shown that the effect of the male is to make her ovaries produce eggs and secrete the hormone progesterone. A female on her own will not lay but will incubate if she has been treated with this hormone. Thus these experiments have shown some of the causes of incubation: in the external world eggs are needed, but the bird must also be 'primed' by the male's courting so that her hormonal state is right as well.

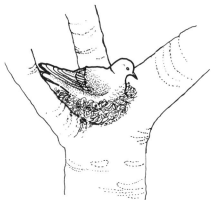

external world so that they can react appropriately to many different sorts of stimuli, escaping from those that are potentially dangerous, attempting to capture and eat those that look like food and approaching and courting those that may be prospective mates. The sensory processes involved are of as much interest to the ethologist as they are to the sensory physiologist and the perceptual psychologist. All have the common aim of understanding how events in the outside world are translated into nervous signals and hence into behaviour.

Both physiological psychologists and ethologists may also be interested in how behaviour is influenced by internal events, such as low blood glucose or high levels of a hormone. Just as it is possible to present an animal with a loudspeaker playing a courtship call or a model showing a threat display, so its internal state can be changed: for example, it can be deprived of food so that its blood glucose is lowered or it can be treated with a sex hormone to see whether this affects its behaviour. A full understanding of the causation of behaviour requires knowledge of how both external and internal events affect the nervous system to produce behaviour. It also requires an understanding of the exact neural mechanisms involved, i.e. the centres and pathways which intervene between senses and movement. This is the realm of the neurobiologist, but ethologists who are interested in causes most often study them treating the animal as a 'black box' rather than probing inside it. For example, they may study the events in the outside world that lead to the performance of a behaviour pattern, or how an animal decides which, of the many behaviour patterns it may perform, it will carry out at a particular moment. The fact that animals are more willing at some times than at others to perform particular responses, like eating, drinking or mating, has led to many different theories of how internal and external factors combine to affect behaviour, collectively known as theories of 'motivation'. Much of the attention of the early ethologists, such as Lorenz and Tinbergen, was devoted to these theories, but recently they have rather fallen from vogue. This is partly because it has become more fashionable to explain the mechanisms underlying behaviour in terms of the animal's neural machinery, as many neurobiologists are trying to do. But it is also because motivation is a very complex matter which requires the taking into account of many different factors for each system of behaviour and cannot be summed up by a simple overall model of the sort that ethologists originally put forward.

1.3 Evolution and function

As mentioned earlier, investigating the evolution of behaviour and its function, or adaptive significance, are two other fields in which ethologists are active, but here their interest is not often paralleled by that of psychologists. In these fields, ethology borders on genetics, evolutionary biology and ecology, rather than on topics of interest to psychologists and neurobiologists. The study of behavioural evolution is not an easy task, for behaviour leaves no fossils, so the best that can be done is to reconstruct the course that it must have taken from a comparison of the behaviour of species alive today. Selection experiments and the study of mutants that behave differently from normal animals can also help us to understand the changes in behaviour that must have taken place during the course of evolution.

The major thrust of recent ethological research has, however, been in the field of function: studies aimed at understanding the adaptive significance of behaviour. The last 20 years have seen a revolution in our understanding of evolution theory, with the emergence of many stimulating new ideas. Some of these concepts, such as 'inclusive fitness' and 'evolutionarily stable strategies', have particular relevance to behaviour and have led to intensive study, especially of the social behaviour of animals in the wild, in an effort to discover just how natural selection has led their behaviour to be as it is. This field of study, on the border between ethology, ecology and evolution theory, is often referred to as sociobiology or, more appropriately as it concerns more than just social activities, behavioural ecology. It has generated some beautiful work, but also a great deal of controversy, much of it rather fruitless. One area of argument has been over the relevance of these ideas to humans, as some sociobiologists are enthusiastic about the application of evolutionary ideas to man, believing that our behaviour can be better understood if viewed in the context of our evolutionary heritage. Another, related, bone of contention has been whether or not the claim that behaviour is adaptive means that it is unmodifiable and under tight genetic control: some sociobiologists have written as if it did mean this, and have hence been charged with 'genetic determinism', an especially heinous crime if they are also writing about humans! But, in truth, all that is required for behaviour to be acted on by natural selection is that it has some genetic basis, no matter how slight. There is no reason why its development should be in any sense tightly controlled, so adaptation does not presuppose genetic determinism. And, in the case of humans, behaviour may come to be adaptive for a great

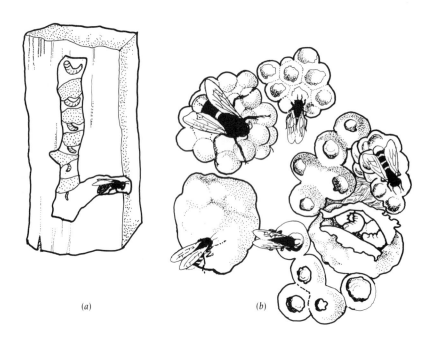

(a)   (b)

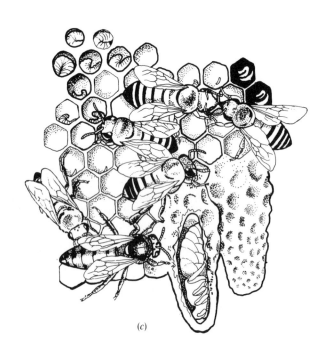

(c)

many reasons other than through natural selection. Indeed, for many of us the environment we occupy is so odd compared with that in which we evolved that it would be hard to argue that our adaptation to it had much to do with selection!

These controversies have generated a good deal of heat, and there are still those who are fiercely committed to one side or the other without a hint of compromise. But reason, as so often, takes a middle course, and neither extreme position is very plausible. Both evolutionary heritage and the genes which are its legacy from one generation to the next clearly influence all the behaviour of all animals, ourselves included, but to argue that either is all-important is as mistaken as to deny that either has any importance. A more serious difficulty with functional reasoning, as with much evolutionary thought, is the ease with which hypotheses can be generated but the difficulty of subjecting them to any rigorous test. Writing popular books packed with brilliant hypotheses dressed up as facts is quite a money spinner for ethologists. Many of the 'Just So Stories' with which these are filled are so compelling that one feels they just must be true – until someone else comes up with an even neater explanation!

The problem is that functional questions are not often open to experiment. To carry out an experiment, one needs two groups of animals, an experimental and a control, which differ only in that aspect of their behaviour the significance of which one wants to test. It is not easy to achieve such a very specific change, leaving all other aspects of behaviour the same. Furthermore, adaptation is only relevant to the particular

Figure 1.7. Studies of evolution often involve comparisons between closely related species. It is not easy to see how something as complicated as the social life of honey-bees may have evolved. By studying the behaviour of the many bee species found today von Frisch was able to suggest a likely course of evolution from a solitary ancestor, through various intermediates, to the social structure of the hive. The bee species shown in (*a*) is solitary, laying its eggs singly in cells which it provisions with honey cake. The young emerge from their cells after the mother has left. In other species, the young hatch while their mother is still laying eggs. They then lay eggs themselves and all combine to tend the young. In bumble-bees (*b*) the first young are small and less fertile than their mother: they produce few young of their own but help their mother rear her later offspring. Finally (*c*) comes the honey-bee, in which only the queen lays eggs while most of her daughters become workers, and devote their lives to tending their sisters and brothers. Thus, by comparing species, we can see how a series of slight differences may, in the course of evolution, have led to the change from solitary life to a social system as complex as that of the honey-bee.

environment in which the species evolved, so the study is most likely to be useful if it is carried out in nature, which further limits the sorts of experiments that can be done. Faced with such problems, many behavioural ecologists have rejected experimentation and concentrated on observation and correlation instead. But this approach has problems of its own. The size of monkey troops may correlate with the size of their home ranges, but this fact does not, in itself, provide an explanation. A larger group may require a larger range to feed on, a larger range may require a larger group to defend it, or both features may be caused by a third factor and not directly related to each other at all. For instance, a food supply which is briefly abundant in a small area, then dies out and 'blossoms' once more some distance away, may lead to both. It may encourage a large home range so that there is always at least one source of food present in it. Furthermore, as each source in turn is abundant, group size can be large without disadvantage to the members of the group. Thus large group and large range will both arise without one causing the other. The moral is that one

Figure 1.8. Examining the function of a behaviour pattern involves trying to discover how it benefits individuals that show it. The cichlid fish *Haplochromis burtoni* has orange spots on the anal fin, close to the cloaca, which are especially bright in the male. These fish are mouthbrooders and, when a pair spawns, the female picks up the eggs in her mouth. The spots on the male's fins are very similar in size and colour to eggs and she often tries to take them into her mouth too. As a result she picks up the male's sperm and ensures that her eggs are fertilised. From this observation we can see the most likely function of the orange spots on his fins. (Redrawn from W. Wickler (1968), *Mimicry in Plants and Animals*, McGraw-Hill, New York.)

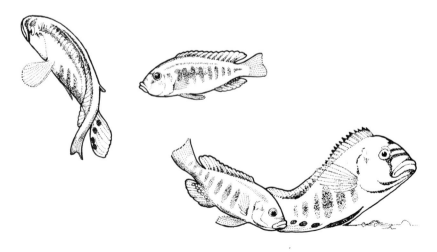

must be cautious about interpreting correlations and that, in function as in other aspects of behaviour, greater certainty can be reached if it is possible to carry out experiments. This is not easy but, with ingenuity, some clever tests of functional hypotheses have been devised.

1.4 Book plan

Ethology, then, is the biological study of behaviour. It concentrates on four different fields of study: causation, development, evolution and function. These four topics will form the core of this book and are dealt with, each in turn, in Chapters 4–7. But the best ethological studies have always started with a period of description, the ethologist getting thoroughly familiar with his animals to be sure that the questions he asks are appropriate. We will follow this tradition in the next two chapters by describing the motor patterns of animals, their form and what is known of their control, then the senses which give the animal its outlook onto its environment, and how stimuli influence these to produce behaviour. Much of the book will be concerned with individual animals, but we will often refer to their interactions with one another, as when they fight or they mate. In the last two chapters of the book, we will come to the social aspects of behaviour more specifically, to discuss communication and the social organisations which are built up by the communicative interactions between individual animals. The literature list at the end of the book is not intended to be exhaustive, but lists a few books in which the reader will be able to take the topics of each chapter to greater depth.

2

Patterns of movement

The ways in which we define the behaviour patterns of animals depend, almost entirely, on the exact movements involved. These movements vary enormously in their complexity, from the slow sweep of a chameleon's eye, which involves just a few very specific muscles, to the headlong dash of a cheetah, in which most of the muscles in its body have some role to play. The simplest of movements, usually referred to as reflexes, may just involve a few sensory cells connected through two or three nerve cells to a few muscle fibres, so they make a good starting point in thinking about behaviour.

2.1 Reflexes

We all know about several of our own reflexes: the blink of the eyelids which occurs involuntarily when something flies towards one's face, the constriction of the iris caused by the flash of a bright light, the knee-jerk reflex induced by a tap just beneath the knee cap. All these require a stimulus from the outside world, but it is a simple one and they are otherwise quite automatic. The knee-jerk, for instance, involves only two nerve cells (Figure 2.1). The connection between these is far down in the spinal cord (though admittedly that is quite a long way up from the knee), but the point is that the brain need know nothing of the action, for it is all carried out involuntarily and does not require any complex processing. This is even so where a third nerve cell, or interneurone, occurs between the sensory and motor nerves, as often happens in reflexes: the interneurone is short and all the connections are far away from the brain itself.

One of the first people to study reflexes was Sir Charles Sherrington, the great English physiologist, who examined various reflexes of dogs in the early years of this century. For much of his work he concentrated on the 'scratch reflex', which dogs use to remove irritation from their sides with their hind paws, and he studied this using the simplest of apparatus. The dog's leg was attached by a thread to a needle which drew a trace on a rotating smoked drum. If the paw was still, the trace was made horizontally as the drum went round but, if the dog scratched, the needle made a line up and down on the drum as it did so (Figure 2.2).

Many of the properties of reflexes Sherrington found also apply to more complex actions, but the fact that they occur even in the simplest acts points

Figure 2.1. The knee-jerk reflex depends on sense organs in the patellar tendon just below the knee cap and on the quadriceps muscle in the upper leg (*a*). Though close, these are not simply connected to one another. The sensory nerve travels up to the spinal cord in the rump region and there synapses with the motor nerve which passes back down to the leg to the muscle (*b*).

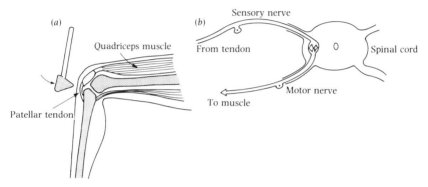

Figure 2.2. The scratch reflex of a dog as studied by Sherrington. The sample trace from one of his experiments shown here illustrates three features of reflexes that he demonstrated: latency, warm-up and after discharge. (After C. S. Sherrington (1906), *The Integrative Action of the Nervous System*, Cambridge University Press, London.)

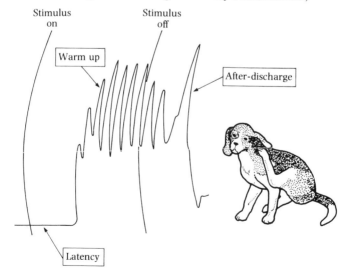

to them being fundamental features of behaviour. Two of the most obvious of these characteristics are latency and after-discharge: when the dog's flank is touched there is a short delay before scratching starts and, when the stimulus is removed, again there is a slight gap before the dog takes away its paw. These are simply explained: it takes time for the nervous signal to travel from the touch receptors in the skin to the muscles responsible for scratching. Signals travel very fast along nerves, but there is a greater delay where they must cross from one nerve cell to the next within the spinal cord. As a result the dog is always a little out-of-date in its behaviour and scratching does not start for a short while after the irritation arrives nor cease till some time after it has gone.

Two less obviously explained properties of reflexes are 'warm up' and 'fatigue', but these two are also important and have their counterparts in other aspects of behaviour. Warm up refers to the fact that the behaviour, once started, does not reach maximum intensity straight away. In the case of the scratch reflex, the first few strokes with the paw do not have such a broad sweep as later ones, as it takes a little time for the action to reach full amplitude, probably as more and more muscle fibres are recruited. In an equivalent way, the rate of alarm calling of a small bird that has spotted a hawk rises over a minute or two rather than being at a maximum as soon as the hawk appears, and an eating mouse nibbles slowly at first and then faster and faster as its meal progresses.

Fatigue is also a common occurrence, in reflexes as in other behaviour. In the case of the dog, it can be seen quite simply by touching the animal's side with something which cannot be removed by scratching. The dog will start to scratch and, after a period of warm up, the intensity of scratching will reach a peak, but it will not persist indefinitely. Slowly it dies away until the animal scratches no more despite the fact that the stimulus is just the same as it was before it began. Something in the system is clearly fatigued, and needs a little time to recover before scratching can start again.

Fatigue is just one aspect of changing motivation: the fact that animals are not always equally prepared to show a particular behaviour pattern when the stimulus appropriate to it appears. In the case of fatigue this is for the specific reason that the behaviour has been elicited a great deal in the immediately preceding period, but there are many other reasons why animals may fail to respond and this wider topic is one to which we shall return in Chapter 4. In the case of reflexes, however, the willingness of animals to perform usually varies rather little provided that they are not stimulated too often. If they are, fatigue does set in, and there are several reasons why it may do so. An obvious one might be that the muscles involved in the action are simply exhausted and incapable of any further exertion. Another possibility is that the sense organs are tired out and are no longer capable of detecting the stimulus. The dog may stop scratching because it does not feel the irritation any more.

These are the simplest suggestions to account for fatigue, but it is often possible to discount both of them because the animal's senses and its muscles are used perfectly normally in other activities while it remains incapable of showing the action which has fatigued. The reaction of a fly maggot to light, studied by Mario Zanforlin, is a good example. After they have finished feeding, these larvae search restlessly for a dark and enclosed space in which to pupate safely. If a light is shone onto them from one direction they will turn and walk away, but what do they do if they then find themselves in a blind alley? Amazingly, as Figure 2.3 shows, after searching around for a while at the end of it, they walk straight out of it again towards the light and then turn away once more when they emerge: a striking example of apparently intelligent behaviour by an animal no one would accuse of brilliance! Could it be sensory adaptation? Perhaps the animal ceases to see the light when in the alley but turns away on emergence because the light is much brighter at the mouth than at the far end. This is not the answer: the maggot will react in the same way even if the light is slowly withdrawn as it emerges so that the light intensity remains the same. Furthermore, maggots free to walk away from light indefinitely will do so for several kilometres without once moving towards it. Nor is the result simply due to muscular fatigue, as the animal remains fully capable of all its other reactions which use the same muscles. Instead the larva is experiencing a very specific deficit: it is incapable of making sharp turns. In a narrow alley it must make these if it is to stay at the end, but after a few of them its turning reaction becomes exhausted and thus the only direction in which it can walk is straight out of the alley. It then encounters a wide open space in which it can make a broad sweeping turn to take it away from the light again. But it will be some minutes before its nervous system recovers enough to allow it to make any more of those

Figure 2.3. *Sarcophaga barbata* larvae move away from light but, if this leads them into a blind alley, they escape from it after a few minutes and then carry on the way they were going.

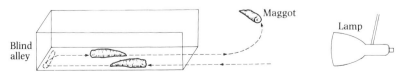

tight turns that it first made when it reached the end of the alley. The evidence is against its fatigue being sensory or muscular, but the incapacity probably arises because certain nervous pathways become unable to carry signals, perhaps through depletion of the transmitter on which they depend for contact between one cell and the next. The pathways need a rest before they can be used again, and during that time the animal cannot show actions that depend upon them.

2.2 Fixed action patterns

Although illustrated with simple reflex actions, these points are equally applicable to more complex behaviour patterns, such as the courtship and threat postures of sticklebacks referred to in the last chapter. Actions such as these have traditionally been called fixed action patterns (or FAPs) by ethologists. There is no hard and fast line between them and reflexes, but they do tend to be rather more complicated and to vary in whether or not they can be elicited in ways that cannot be put down to simple 'fatigue'. Many of them are only produced in response to quite complicated stimuli: for example, size, shape, hue and texture may all contribute to the ability of a visual stimulus to effect a response. This is a topic to which we shall return in the next chapter: it is enough to point out here that this is a far cry from the sharp tap or the change in light intensity which may be the critical stimulus for a reflex. Integrating such complex information from the outside world obviously also involves more than a few neurones: processing in the sensory areas of the central nervous system may be involved in the decision of whether or not action is taken. And the action itself, no matter how stereotyped, will often involve hundreds of different muscles all contracted in well ordered sequences, again involving many neurones.

The idea that animals possessed a repertoire of FAPs originated with Konrad Lorenz. He described their properties very clearly and this makes a useful starting point in considering how ideas have changed in the intervening half century. One of his claims was that these acts are innate. But, as pointed out in the last chapter and discussed in more detail in Chapter 5, this is not a very useful criterion, as all behaviour depends on the animal's inheritance, so no sharp line can be drawn between actions which are innate and those which are not. Another suggestion was that FAPs are 'invariant', showing negligible differences between different individuals or between repetitions by the same individual. Indeed this is implied by the word 'fixed'. But detailed analysis shows that this is not as true as it seems at first sight. Some actions are certainly very stereotyped, and this is especially true of many courtship signals for the simple reason that one of their main functions is to indicate to prospective mates the species to which the individual belongs. A well-studied example here is the head-toss of the male goldeneye duck (Figure 2.4). Woe betide the bird whose head-toss diverges from the norm, for he will get no mate. The average head-toss is 1.29 seconds long, and the standard deviation, which is a conventional measure of variability, is only 0.08 seconds. What this means is that 95% of tosses are between 1.13 and 1.45 seconds in length, which is indeed pretty constant. But not all actions are as fixed as this and it would be artificial to draw a line, calling some of them FAPs and some not. It is best just to bear in mind that quite a variety of different actions are lumped under this heading, some more fixed than others. This may appear confusing, but the fact that animals are not automata is half the fun of the subject! Nevertheless, it is true that many animal species do have an array of stereotyped and species-typical patterns of behaviour. Some may only be shown by males in breeding condition, some only by infants, some by lactating mothers, but within each category all animals will show them in more or less the same form.

Some contrasts between species highlight the way in which FAPs are the same throughout a species but may be different between them. When they walk, most mammals do so with their diagonally opposite legs off the ground at the same time, which undoubtedly helps to keep their centre of gravity above the two that are in contact. But, perhaps surprisingly for an animal with its centre of gravity quite so high, the giraffe does not (Figure 2.5). It raises both left legs, then both right ones, swinging its body from side to side as it goes. This falls neatly into the category of a stereotyped and species-typical aspect of behaviour: all horses walk one way, all giraffes the other. Drinking in birds provides another contrast. Most birds

Figure 2.4. The head-toss display of the drake goldeneye. Like many duck displays this is very stereotyped, the male suddenly tossing his head back and then bringing it forward again in almost exactly the same way every time.

cannot swallow water with their heads down but take a sip and then raise
their heads to let the water slide down their throats. Pigeons, on the other
hand, do not have this problem but simply suck the water down. Again
this is a hard and fast rule: one will not find the odd sparrow that has
mastered the art – nor the odd pigeon that has not. These distinctions are
sometimes useful to taxonomists, in their efforts to classify species of
animal into groups, for behaviour can be as much a species-typical
characteristic as can number of toes or coat colour. Among wading birds,
one of the reasons why the oystercatcher is classified as closely related
to the plovers depends on its behaviour. Like a plover it scratches its head
by raising the leg over the top of the wing, whereas other waders bring
the leg up under the wing.

    These examples illustrate the fact that closely related species may often
share patterns of behaviour with each other. Similarities between species
are most striking in the sorts of actions just described, such as walking,
drinking or head scratching. They are less often true of courtship acts for,
just as the male goldeneye must get it right to ensure he is accepted by

Figure 2.5. When walking, giraffes raise the two legs on one side of
the body at the same time, whereas other mammals, such as horses,
show a 'diagonal' walking pattern, with legs at opposite corners of
the body being lifted simultaneously.

a female of his species, his courtship must be sufficiently different from that of other species of duck to avoid attracting their females. To return to sticklebacks for a moment, the three-spined and the ten-spined both show zig-zag dance in their courtship, but the female does not have to resort to counting spines to decide on the male with which she should mate! The ten-spined is jet black where the three-spined is red, and his dance is carried out zig-zagging vertically with his head down towards the mud, as if he was hopping along on a pogo stick, rather than to and fro horizontally. Thus females will have no difficulty in selecting a mate from the right species.

Rather than stressing the innateness and invariance of FAPs, ethologists today tend to think of each species as having an array of relatively stereotyped behaviour patterns which will be shown by all its members, at least of the same age and sex. Some may call them FAPs, others simply refer to them as behaviour patterns or acts: these are certainly descriptions which avoid any implications to which anyone might take exception! Subsequent studies have shown that few, if any, behaviour patterns would fit in with the defining characteristics that Lorenz laid down when he first described FAPs. One of these illustrates this rather well: his suggestion was that FAPs were triggered by stimuli in the external world but, once started, became independent of them. In proposing this, he no doubt had in mind the egg-rolling behaviour of the greylag goose, which he and Tinbergen studied at around the time he first described FAPs.

If an egg rolls from its nest, the goose leans out, places its beak beyond the egg, and then carefully draws the beak back towards its chest so that the egg is retrieved (Figure 2.6). Lorenz referred to this action as having

Figure 2.6. The egg-rolling response that a greylag goose shows when an egg has rolled from its nest. (Redrawn from K. Z. Lorenz & N. Tinbergen (1939), Z. *Tierpsychol.* **2**, 1–29.)

a fixed component, the movement of the beak towards the chest, and a variable component, side to side movements of the beak which serve to keep the egg in place. The latter are modified by the movements of the egg but the former, Lorenz argued, is independent of the egg once started and so constitutes the truly fixed part of the action pattern. As a demonstration of this, he and Tinbergen removed eggs from geese which were in the middle of rolling them back: a somewhat risky procedure, as anyone who has attempted to approach an incubating goose will realise! Remarkably, the animals continued with the action till their beaks returned to the nest, even though the original stimulus had been removed. The action thus appeared to be triggered by the egg but not to depend on its presence thereafter.

There are not many other examples of behaviour patterns which run to completion regardless of information from the outside word. Most of them, as we shall see in Chapter 4, are strongly dependent on feedback from what they achieve. The contrast between prey capture in two species will highlight the difference between behaviour patterns which rely on such feedback and those those which do not (Figure 2.7). If a dog sees a rabbit, it runs towards it watching it all the while, so that if the rabbit moves off it can change its trajectory to maximise the chances that their paths will meet. Generally speaking dogs do not arrive at the spot where a rabbit originally sat and wonder where it went to! Their movements are continually changed in the light of incoming information or, in more general terms, in the light of feedback from the results of their actions. A praying mantis, on the other hand, if it finds a fly within striking distance, will rock to and fro and then suddenly pounce with alarming rapidity. Should the fly move when the mantis has initiated the pounce, its meal is lost, for it cannot change the course of its strike once it has

Figure 2.7. A dog chasing a rabbit changes course the whole time in response to the rabbit's movements and thus achieves interception. By contrast, if the prey of a mantis moves in mid-pounce, the mantis cannot react but still strikes at the spot where the fly was when the strike began.

started. Like the goose, this is an example of an action which fits in with Lorenz's description: it is only triggered by the external world but not modified by it as it proceeds, so that once initiated it goes to completion regardless. In the case of the mantis, it is easy to see why: the action is so rapid that there simply is not time for feedback – nor indeed is there much time for the fly to move beyond the area of its strike. But most behaviour is less rushed than this, and is modified by feedback from its consequences, so that few acts would be defined as FAPs according to Lorenz's strict definition.

2.3 Central and peripheral control

A question related to whether animals sense changes in the external world and modify their behaviour in line with them is whether internal changes brought about by their actions affect how they behave. To some extent this must obviously be so. The more one eats the more satiated one becomes and clearly satiation results from sensing that one has had enough: such factors as having a full stomach and high levels of nutrients in the blood contribute to this feeling. But consider a shorter term action like the sequence of leg movements in walking. This could be organised in two quite distinct ways, one relying on peripheral information and the other not. The brain might send out detailed instructions to each of the relevant muscle blocks in turn, patterned in the right sequence to bring about the relevant leg movements. This would not need any knowledge of what the legs were actually doing. At the other extreme, the brain might simply send a message to the first muscle to contract, this muscle might be linked to the next, and so on until the last muscle in the sequence was linked back to the first again. With such a system, cycles of walking would occur because of the connections from one muscle to the next, the only instruction from the brain itself being the order to start and to stop.

These are two extreme ways in which movements might be arranged, the one where patterning was central, the other where it lay in connections at the periphery. In practice, motor organisation tends to rely to some extent on both sorts of influence. The sequence of muscular contractions in walking does appear to depend upon central instructions, but its details can be modified by information from sense organs at the periphery. An example of such a peripheral influence is the crossed-extensor reflex which has the effect, when one leg is raised from the ground, of causing contraction in the extensor muscles of its neighbour, so leading it to stiffen and bear the weight of the body. Such actions will obviously help in the

Figure 2.8. Feedback enables a cross-country runner (*left*) to move over rough ground, modifying the tension in different muscles and shifting his balance to counteract the effect of the bumps and tussocks he encounters. To achieve similar feats, a robot (*right*) would have to be very complex: most can only cope with smooth paths as they have no feedback system to enable them to respond to rough patches of land. (Photographs by N. Luckhurst.)

adjustments necessary during movements. Similarly, information from the limbs to indicate exactly when they have hit the ground, how much resistance they are receiving from it and how much weight they are bearing, will be essential to control the fine timing of the sequence of actions. Imagine a cross-country runner moving over rough terrain (Figure 2.8). In such a situation a robot from a science fiction film simply following the instructions from its computer without any feedback from what its actions achieved would rapidly fall over. It is no coincidence that such robots are always filmed moving behind some rubble so that one cannot see the smooth path that has been specially laid for them!

These points illustrate the importance of peripheral modulation to adjust movement in keeping with environmental changes. But modulation is, for most movements, as much as the periphery provides, the basic pattern being passed down from the brain. This enables changes of speed and of gait to be imposed from above and it also enables the same muscles to be used for many different actions. If a stimulus to one muscle set in train a series of commands to others so that the animal started walking, the same muscle could not be used for kicking or for scratching, at least without complex connections to inhibit the unwanted sequences which would otherwise follow automatically from its activity. The combination of central patterning and fine feedback control from the periphery assures the utmost precision of movement; one need only watch an Olympic gymnast to be impressed by what such an arrangement can achieve.

The central origin of motor patterning has been most clearly shown in some invertebrate systems. An especially good example comes from silk moth eclosion: the movements shown by the moth as it twists and turns to free itself and emerge from its pupa (Figure 2.9). This is not what one might at first sight call a fixed action pattern, as the whole procedure takes about one and a half hours to complete, but it is stereotyped and species-typical, so it fits in well with our definition. Such a long drawn out procedure also has the advantage of being leisurely enough to be studied without slow motion equipment! The sequence is split into three half-hour

Figure 2.9. Recordings from the motor nerves in the isolated abdominal nervous system of a silkmoth after eclosion hormone has been added. In (a) the ganglia are shown with left (L) and right (R) motor nerves labelled. At first all the nerves on the left fire alternately with all those on the right (b). In the traces shown, R4, R5 and R6 are firing simultaneously, but in alternation with L4. Then, after a period of quiescence, waves of firing pass up the body (c), from R6 to R5 to R4, and now the motor nerves on the two sides of each segment fire simultaneously rather than in alternation. Thus L6 and R6 show peaks at the same time in the example shown. (After J. W. Truman & J. C. Weeks (1983), *Symp. Soc. exp. Biol.* 37, 223–41.)

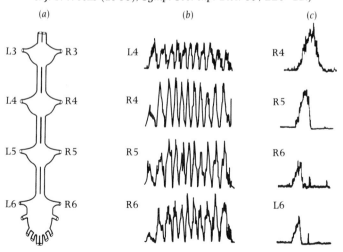

periods. In the first (Figure 2.9*b*), all the muscles down one side of the body contract alternately with those down the other, so that the abdomen twitches from side to side and becomes detached from the pupal case. In the next half-hour the animal rests quietly and in the last, it performs a quite different sequence of muscular movements which lead it to emerge from the pupal case. During this period waves of muscular contraction pass from segment to segment down the body, the muscles on the two sides of a segment now contracting at the same time rather than alternately (Figure 2.9*c*). These peristaltic movements, like those of an earthworm moving through its burrow, have the effect of moving the case off the end of the body, so freeing the moth.

All of these very precise movements are patterned entirely within the four abdominal ganglia of the moth and require no sensory connections whatsoever. If these ganglia are carefully dissected out from the moth, retaining their connections with each other but cutting the motor and sensory nerves leaving them, recordings can be made from the remains of the motor nerves to see what the ganglia on their own can produce. Amazingly, the same sequence is found as in the intact moth. For about half an hour all the nerves on one side fire in alternation with those on the other, then there is a period of quiescence, and then the nerves to both sides fire from each ganglion in turn. The pattern is just the same as in the intact moth but there are no muscles to translate it into action. What then stimulates this sequence if sensory nerves play no role and nor do connections from the ganglia further forward in the body? A trick is needed to get the program started: the sequence will not get under way unless the appropriate 'eclosion hormone' is added to the medium in which the ganglia are kept. This hormone normally starts to circulate in the body when eclosion is due and sets the whole process in motion.

2.4 Orientation of movements

The eclosion of silk moths is a remarkable example of a behaviour sequence which is complex and yet entirely organised within the nervous system without the need either for sensory influences from outside the animal to trigger it or for sensory feedback to affect its course. This is very unusual, for in most behaviour sensory information has a crucial part to play, even if only to ensure that the movements of the animal are correctly orientated with respect to the outside world. Here an even longer term series of actions, the migration of European warblers, serves as a good example. These small birds fly south in the autumn, some species only going as far as the Mediterranean, but others travelling on over the Sahara into the southern half of Africa. Sensing the correct direction to take depends on outside cues, such as the stars and the earth's magnetic field, but other aspects of this migratory pattern occur regardless of any input from the outside. Even when kept in laboratory cages with constant daylength, the birds become restless at night twice a year in spring and in autumn at the times when they would normally migrate. The species which migrate furthest are restless for more nights than those which travel a shorter distance, and the number of nights matches that which would be required to reach their destination. The birds will even flutter towards the end of their cage nearest their direction of travel. For garden warblers, which winter far down into Africa, this direction is south-west at first and then south after they have crossed the Sahara so that they avoid flying out into the Atlantic. Once again this change is found also in the caged birds (Figure 2.10): each autumn their fluttering takes on a less westerly component towards the end of the series of nights on which they are restless. This is true even of hand-raised birds which have not themselves ever been outside Europe.

The egg-rolling of a goose must be oriented in relation to the egg, and the migration of a warbler must carry it in the correct direction over the surface of the earth. But, beyond their role in orientation, external stimuli play little part in the pattern of these movements. The next chapter will consider more complex external stimuli and the role they play in behaviour, but here it is worth staying with orientational mechanisms for a little longer as these fall into various different categories which are quite helpful in our thinking about movement patterns. Indeed, before ethology became widely popular, one of the main theories of behaviour was based upon classifying it by the various different forms of orientation involved.

The simplest of these mechanisms are called kineses, and these do not even involve the animal sensing the direction of the stimulus. If one side of a dish of water with some flatworms in it is covered with a dark cloth, the animals will be found to accumulate in the darkened area (Figure 2.11). But they do not do this because they are escaping from the light: it is because they move more slowly and turn more often in the dark. This is a kinesis because the direction of the light makes no difference, while movements such as that of the fly larva which walks away from a source of light are referred to as taxes. These take various forms, the simplest being those directly towards or away from a source of stimulation, whatever this may be. Positive phototaxis takes an animal towards light; negative geotaxis away from the earth's gravitational pull.

Figure 2.10. The migratory pattern of garden warblers which nest in
Europe and then move to southern Africa in winter. The arrows on
the map show the preferred directions of hand-reared birds kept in the
laboratory and tested on the dates when free-living ones would be at
the locations shown. As autumn progresses, their preference moves
from south-west to south in keeping with the route they would follow
in the wild. (After E. Gwinner & W. Wiltschko (1978), *J. comp. Physiol.*
125, 267–73.)

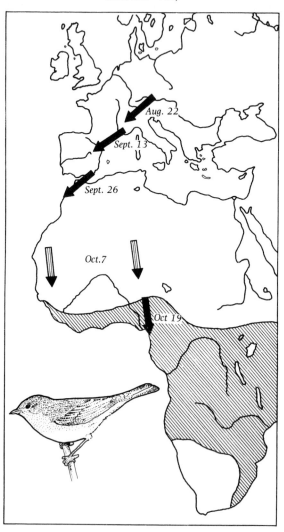

One stage more complicated than this is a movement which is at an angle to the stimulus rather than straight towards or away from it. Homing pigeons sometimes use the sun to tell them the compass direction of home, they then fly at the angle to it appropriate to reaching their destination. This is not just a matter of taking up an angle, however, for the sun moves through the sky and they must compensate for this movement if they are to keep a constant direction. If they are in the northern hemisphere and home is south, the sun should be to their left early in the morning, ahead at noon and over to the right at dusk. In other words, they must know what time it is as well as the direction in which they wish to travel: a not inconsiderable feat. In many ways it is rather easier for the warblers which migrate down to Africa, for they fly almost entirely at night. One outside cue that they use to guide them is the pattern of stars in the sky, and especially the small cluster of stars round Polaris, the pole star. These are useful because the star pattern rotates round them and they remain in the same position regardless of the season or the time of night. A bird wishing to fly south need only fly away from them and does not require to know what time it is. Small birds also use the earth's magnetic field, in ways that are not yet full understood, especially as far as the sensory mechanisms involved are concerned. An interesting example here is the European robin, which uses the angle of dip of the field to detect which direction is north

Figure 2.11. Flatworms (*Dendrocoelum* sp.) in a half-covered tank gather in the darkened area. This is not because they avoid light but because they move more slowly and turn more often when in the dark.

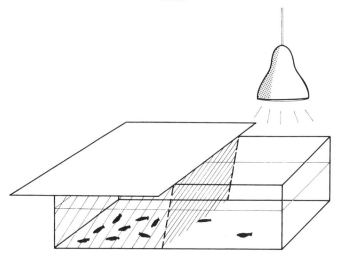

and which south (Figure 2.12). A compass needle held vertically points horizontal to the ground at the equator and straight down at the north pole. As robins do not cross the equator they can be sure, without even knowing the direction of the field, where north is because the lines of force point downwards towards it. Perhaps it may seem curious that they do not have to know the polarity of the field, in other words which end of a compass needle is which. However, a good reason for this is that which of the earth's poles is which has switched every so often during geological time: though compass needles point north today, there have been periods when they would have pointed south. Thus, if robins depended on polarity, and a switch occurred they might all fly off to the north pole in autumn!

A final form of orientational mechanism is the way in which animals use landmarks to find their way around. This involves sensing the complicated patterns and relationships of objects in the world around them rather than just the direction of this or that stimulus. It thus involves sophisticated sensory systems and brain mechanisms such as will be discussed in the next chapter. The navigation of robins or of homing

Figure 2.12. In spring, European robins move north and they find this direction by taking the smallest angle between the earth's magnetic field and gravity (γ in *a*). That they do this rather than using the polarity of the field has been shown by experiments in which the field direction is reversed (*b*). Here the birds still orientate by the smallest angle, but they now go the opposite way from that the direction of the field would indicate. (After W. Wiltschko & R. Wiltschko (1972), *Science* **176**, 62–4.)

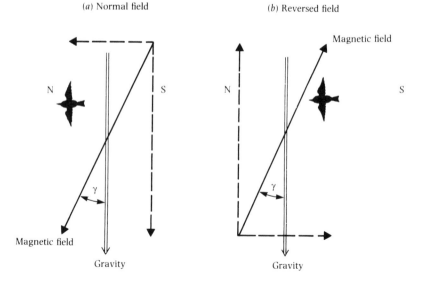

(*a*) Normal field   (*b*) Reversed field

pigeons may get them close to their destination but could hardly land them in the same bush in which they nested last year or on the roof of their loft from hundreds of kilometres away. Knowing the features of their home environment undoubtedly helps them to reach the exact place after they have arrived in the rough area using their skills at long distance navigation. Indeed orientation by landmarks is an important capacity in most animals whether or not their movements normally take them far from home. They must move around their range to find food and water, shelter and mates. It is essential that an animal has a detailed knowledge of the places where each of these may be found and how they may be reached from wherever it happens to be. For example, imagine a rabbit feeding in the open when a hawk flies over. Its chances of survival depend crucially on its ability to reach the safety of a burrow within seconds. In some sense it must therefore have a map of its environment in its head and a knowledge of just where it is on the map: not surprisingly it is an important function of the nervous system to build up such maps.

3

Sensory systems

Whether it is during long distance migrations, discussed at the end of the last chapter, or when trying to decide if the sounds made by a courting male belong to the right species, or in tasting food to see if it is palatable or poisonous, animals must keep a constant check on their environment through the medium of their senses. These provide them with windows out onto the world in which they live, but they are windows whose features vary greatly from sense to sense and from species to species.

In humans, the two senses of vision and hearing are particularly finely developed, and we tend to think of these, rather than touch, smell or taste, as our primary contact with the outside. It is certainly marvellous the detail with which they provide us, especially when one considers that the scene at which one is looking and the sounds to which one is listening are not there inside one's head at all, but have been reconstituted from a stream of nerve impulses coming from the eyes or ears. These nerve spikes do not vary in size and shape to carry information, but are simple all-or-nothing signals. Which nerve fibre from the eye they pass along indicates the position, and sometimes also the colour, of the object to which they are responding; the rate at which they are sent indicates how bright it is. Similarly, nerve cells from the ear are stimulated by particular tones, and their rate of firing indicates the degree of loudness. With literally thousands of receptor cells, each tuned in a different manner, and with two ears to help us to detect from whereabouts different sounds come, our brains are presented with all the information to reconstitute the sound, be it Rod Stewart or Rachmaninov. It is then up to our psychology whether we appreciate it or not!

Many other species, especially birds, as their bright colours and varied songs suggest, also rely mainly on sight and sound, but their senses often differ from ours in acuity and in frequency range. For example, barn owls have remarkably acute hearing which enables them to locate sounds much more precisely than we can do (Figure 3.1). Like ourselves, most birds can hear up to around 20 000 cycles per second; the hearing of some animals,

Figure 3.1. A barn owl swoops down to capture a mouse, its movement caught in a series of flashes. The hearing of these birds is extremely acute and they can localise a source of sound in three dimensions so accurately that they have no difficulty in capturing prey in pitch dark. (Photograph by M. Konishi.)

of which bats are the most famous, goes up far beyond this to around 120 000 cycles per second. Bats use this ability mainly for echo-location: they produce high pitched sounds which bounce back off objects and so enable the animal to 'hear' where the objects are. Just as we are unable to hear the 'ultrasounds' that bats use for this purpose, we are also unable to see some of the colours that other animals can sense, those with wavelengths shorter than violet or longer than red. Insects are sensitive to ultraviolet light and can, as a result, be attracted to flowers by colours and patterns on them that we simply cannot see (Figure 3.2).

It is more difficult for us to have any idea of the uses to which animals put the senses which we use rather less. The complex world of smells at the foot of a lamp-post may well be, to a dog, as exciting and full of information as is a magnificent painting or piece of music to us. Our reliance on touch is so slight that it is also hard for us to gain a 'feel' for the experience of an octopus as it wraps all its tentacles around an object to explore its shape. Even more difficult are the cases where animals possess senses we do not have or at least have no awareness of possessing. The rattlesnake uses two pit organs to sense the heat coming from its prey and, by measuring the differences between them as we do those between our ears, can build up a map of the heat around it to locate the prey exactly. Some fish have electric organs with which they set up a field around them like the lines of force around a magnet (Figure 3.3). They use changes that occur in this field to locate objects and they themselves can alter the field as a way of signalling to one another. Lastly, there is the magnetic sense referred to in the last chapter, and now shown in bacteria, bees and birds. In birds and bees, this sense probably relies on organs containing small amounts of magnetic material which are affected by the earth's magnetic field and so can be used by the animal to help it find its position.

The diversity of sensory systems is a fascinating subject in its own right, but our task here is to extract some general principles about how the information they provide to the nervous systems of animals affects behaviour. We will take most of our examples from the senses which have been most studied because they are the ones we ourselves use the most – vision and hearing.

3.1 Releasers and sign stimuli

When a female stickleback in breeding condition enters the territory of a male that has completed his nest, he shows the zig-zag dance, but when another male appears he behaves quite differently, showing the head-down threat posture. How does he tell the difference? In a classic series of experiments, Niko Tinbergen used various models, including those shown in Figure 3.4, to explore this question and found that the most important feature stimulating the zig-zag dance was the swollen belly of a ripe female,

Figure 3.2. To our eyes the patterns on some flowers and insects are not especially striking (*above*), but when photographed under ultraviolet light (*below*) they are seen to be much more elaborate. This is because the function of the patterns is to signal to insects and their vision stretches into the ultraviolet. (After Eisner *et al.* (1969), *Science* **166**, 1172–4.)

and that leading to threat was the red on the underside of the rival male. Sausage-shaped models lacking fins, gills, spines, tail and eyes were responded to with courtship provided that they had a bump beneath like a female ready to spawn. As far as threat was concerned, Tinbergen even described how the males in his aquarium tanks would sometimes show the head-down posture when a red post-office van drove past the window! Clearly, the colour red on its own was enough to give the display.

Experiments like these show that some features of a stimulus may be sufficient to bring about a particular pattern of behaviour. The general term for such features is 'sign stimuli', indicating that the animal attends particularly to them rather than to other details which may, on the face of it, be equally striking. In the stickleback case we can see why this might be so: the critical difference between a male and a female intruder is that the male has a red underside; that between a ripe and an unripe female

Figure 3.3. (*a*) The electric fish *Gymnarchus niloticus*, showing the position of its electric organ in strips spreading forward from the tail. The receptors are largely concentrated in the head. (*b*) The fields of force around *Gymnarchus* as altered by the presence of an object of high conductivity (*left*) and by one of low conductivity (*right*). (After H. W. Lissmann & K. E. Machin (1958), *J. exp. Biol.* **35**, 451–86.)

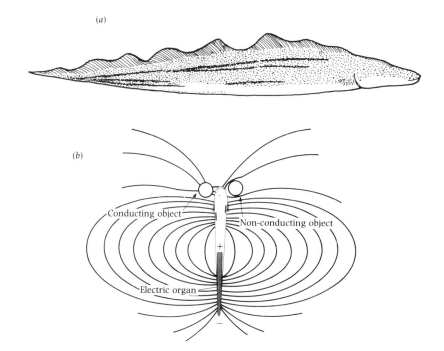

(*a*)

(*b*)

lies in whether her belly is swollen. It seems appropriate, therefore, that natural selection has made the territory owner especially responsive to these points. In some cases there is reason to believe that selection has also changed the feature concerned to make it more effective as a signal. This is probably not the case with female sticklebacks, who are made swollen by eggs anyway. But the red of the male is clearly there because it acts as a signal to other males to keep out and to females to approach.

Features such as this, which have arisen during evolution specially to function as signals from one animal to another, are known as releasers. Examples are numerous, from the colour patterns of butterflies and coral reef fishes, to the songs of birds and of cicadas, and to the smells with which mammals mark their territories and moths attract their mates. Releasers act in communication, and so we will return to them again in Chapter 8, but many of the sign stimuli to which animals respond are not releasers in this strict sense but simply aspects of the outside world which provide valuable cues to them and so alter their behaviour. The pattern of stars in the sky or the sound of a mouse rustling in the grass may influence the behaviour of a hawk, but they are certainly not signals sent to it specially to do so. Strictly speaking these are sign stimuli but not releasers, though the two terms are often used interchangeably partly because it is not easy to tell with some features of an animal whether they are specially adapted to act as signals. Does the belly of the female fish appear to bulge more than the eggs would naturally make it do?

As with fixed action patterns, the early ethologists had some very clear ideas about the nature of sign stimuli, many of which have had to be

Figure 3.4. Three-spined stickleback models as used by Niko Tinbergen. That of a normal male (*a*) and an unripe female (*b*) are very similar, except for the red colour on the underside of the male. The details are not necessary for males to respond appropriately, however, as they will attack a crude model provided this has a red underside (*c*) and court one with a swelling beneath (*d*). (After N. Tinbergen (1951), *The Study of Instinct*. Clarendon Press, Oxford.)

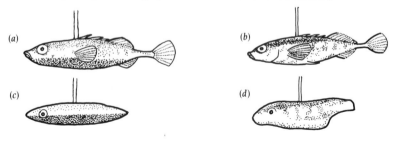

modified as a result of later work. One evocative analogy was with a lock and key mechanism. The sign stimulus was seen as like a key which turned a lock somewhere inside the animal, referred to rather grandly as the 'Innate Releasing Mechanism' (IRM), and so caused the animal to perform the appropriate action. Whether a single mechanism is usually involved in producing such responses, whether it releases the behaviour more or less automatically, and whether it is in any sense inborn, have all been questioned in particular cases by later work, so the expression IRM is not now normally used. But there is, nevertheless, no doubt that some of the ideas it encapsulates are useful. For example, animals certainly do show selective responsiveness, paying more attention to some aspects of a stimulus than to others. But in most cases the response is not simply a lock-and-key one, some features being all-important, the rest totally irrelevant. An example will help to illustrate this point.

3.2 What is an egg to a gull?

The egg of a herring gull is shaped like that of a hen but is rather larger. It tends to be green or brown in colour and to have a number of large dark blotches on it, especially at the blunt end. Like the goose referred to earlier, the gull will retrieve one of its eggs if it rolls out of the nest, and the Dutch ethologist Gerard Baerends has used this behaviour in an ingenious and lengthy series of experiments to determine just what stimuli tell the gull that the object outside its nest is indeed an egg.

What Baerends did was to place two model eggs side by side outside the nest of a gull and watch from a hide to see which of them the gull rolled back first (see Figure 3.5). Over a series of thousands of such tests he varied the characteristics of the models so that he could see what were the preferences of the gulls. To avoid confusion, he changed only one feature at a time. For example, he tested the gulls with eggs varying in size from that of a pigeon's egg to that of an ostrich's, but kept their shape, their colour and their blotchiness the same throughout these tests. When examining the influence of shape he used models in the form of prisms, cylinders and rectangular blocks, as well as ones like eggs, but all of these were kept the same size and painted similarly so that he could see which shape was most effective.

One reason why it was necessary to carry out an enormous number of tests to decide which features were most important to the gulls was that the birds showed a strong position preference, so that with two similar eggs some birds would almost invariably choose the left-hand one and some the

Figure 3.5. A herring gull returns to its nest to find a choice between two egg models on the rim. (From G. P. Baerends & R. H. Drent, eds. (1982), *Behaviour* **82**, 1–416.)

right. To allow for this, each bird had to be presented with many different pairs and their position had to be changed around in a systematic way. When such allowances were made, some very clear results emerged. For example, the birds tended to prefer the larger of two eggs, even if it was much bigger than their own and the other was the normal size. They also preferred egg models with many small speckles to the more natural ones with a few large blotches on them. Both these features probably arise because eggs which are larger and with more spots on them are more striking. Indeed it could simply be that they stimulate the eyes of the bird more and so are more likely to attract its attention.

In the case of other features the preference tended to be for those more like normal gull eggs. The birds preferred models shaped naturally to those in the form of cylinders or rectangular blocks. Amongst prism- and block-shaped models, they chose those with rounded edges rather than ones with sharp corners. They preferred models with blotches darker than the background to those with lighter ones. They chose green, yellow or brown eggs rather than blue, red or grey ones. A slight difference between their colour preference and features of the normal egg is that they retrieved yellow egg models rather than brown ones, though the normal egg is green or brown but never yellow.

All these tests were carried out on models, and these obviously cannot be made to look exactly like the real egg. For instance, the wooden ones Baerends used lacked the textured surface typical of egg-shell. Nevertheless, a model could be made that the gulls would roll back in preference to one of their own eggs. It was 50% larger than normal, green, and with many tiny black speckles on it. This is a good example of what ethologists refer to as a 'supernormal' stimulus as it is more effective than the stimulus found in nature.

These results show that many factors contribute to the egg-rolling response. Though an egg may be retrieved even if its size, shape and colour are wildly different from the gull's own egg, all these features do have a role in making the animal more likely to respond. One feature may be enough, but many others do contribute. At first sight this may seem rather different from the stickleback example, but it is not really. Although male sticklebacks may threaten any red object, without very extensive experiments one cannot determine whether red is all that matters or whether they would threaten more if the object was the right shape, had fins and a tail, spines and an iridescent blue eye.

What both sets of experiments do illustrate is that animals may respond differently to objects that look quite similar to each other and that certain features of a stimulus may be more important than others in leading to their response. We must now consider the sensory basis of these differences and, in particular, the extent to which they have their origin in the responsiveness of the sense organs rather than in the brain. The sensory world of frogs and toads has been studied extensively, and this provides some excellent examples.

3.3 What the frog's eye tells the frog's brain

The title of this section is taken from a classic paper on the vision of amphibians by Lettvin and his co-workers. It might seem obvious that the animal's eye should tell its brain as much as possible about what the world around it looks like, but this is not what it does at all. The eye of a frog or toad, is actually more complicated than ours, with several layers of nerve cells between the light receptors and the nerve fibres which travel from the eye to the brain (Figure 3.6). The ganglion cells are the last staging post before these fibres leave, and the diverse connections of the cell layers

Figure 3.6. The cell organisation in the retina of a primate (*a*) and that of a frog (*b*). Similar classes of cells exist in both cases but, in the primate, bipolar cells connect the receptors directly to the ganglion cells, which send axons to the brain, while the arrangement is more complex in frogs, which have no such direct connections. (Redrawn from C. R. Michael (1969), *Scient. Amer.*, May, **220**, 104–14.)

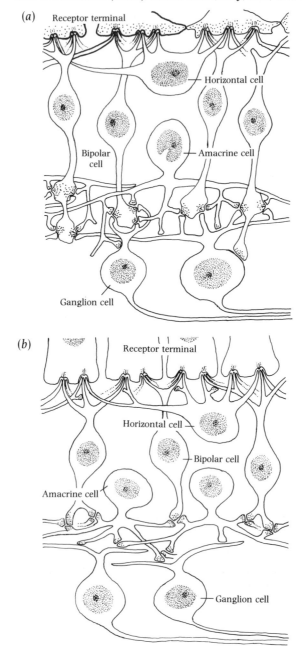

which precede them mean that each ganglion cell receives input from a large number of receptors.

Recording from ganglion cells shows that some of them respond best to very particular objects. For example, one class fires when a small rounded object passes into the field of view and will continue to do so if the object becomes stationary. Interestingly, this cell will not fire if such an object suddenly appears in front of the frog without being seen to move in; nor will the cell respond to a larger dark bar moved across in front of the eye. Because of their particular characteristics these cells in the retina of frogs have been aptly named 'bug detectors', and that is probably their exact function. Objects that cause them to fire are much more likely than other things to attract the attention of the frog, and they are also much more likely to be food than are vertical rods or static spots. Broad edges moving into the frog's field of view, and sudden shadows moving across it, stimulate other types of ganglion cell, and these are probably concerned with recognition of much larger objects such as hawks and other predators.

Thus the frog's eye does not tell the frog's brain about things that are irrelevant to it, but selects only the important features to which the animal should respond, such as stimuli likely to be edible and those likely to be dangerous. To the former the frog will flick out its tongue, which is very long and has a sticky end on which flies become trapped; to the latter the frog will show escape, and here the colour responsiveness of its eye also has a part to play. If faced with a green window and a blue one, side by side (Figure 3.7a), a frightened frog jumps towards the blue on 85% of occasions, perhaps because it is better to jump into the air or towards water than into grass when a predator appears. Again, the physical basis of this response appears to lie in the retina itself: it has been found that cells in the optic nerve fire more in response to blue objects than to green (Figure 3.7b).

Despite the fact that the eye of a frog or toad clearly passes on only the most relevant information to its brain, so that only a very rudimentary picture of the world can be recreated there, not all the animal's reactions can be put down to which types of ganglion cell in its retina have been stimulated. For example, when a toad snaps at prey it must assess the distance away of that prey so that it can extend its tongue by an appropriate amount. To do this requires binocular vision involving integration of information from the two eyes, and this is carried out in a part of the brain called the optic tectum which consists of two nuclei, one on either side (see Figure 3.8). The nerve fibres from each eye travel back to the optic tectum on the other side of the brain and the connections they make there are

Sensory systems

Figure 3.7. (*a*) A frog confronted with two windows towards which it can jump, a green one and a blue one. (*b*) Firing rates of fibres in the frog's optic nerve in response to various colours. (After W. R. A. Muntz (1964), *Scient. Amer.*, March, **210**, 111–19.)

(*a*)

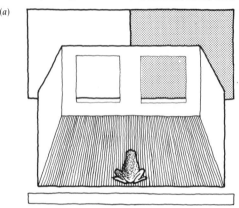

(*b*)

Violet-blue

Blue

Blue-green

Green

Green-yellow

Yellow

Orange

Red

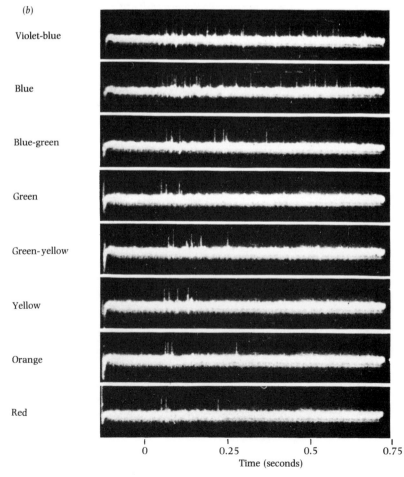

0	0.25	0.5	0.75

Time (seconds)

mapped out over the surface of the tectum in the same arrangement as are the positions in the eye from which they came. But each cell in the tectum is also connected across to another cell seeing the same point in space on the other side, as shown in Figure 3.8. These binocular cells can be used by the animal to assess distances.

The role of the brain in the selective responsiveness of amphibians has been shown particularly well by work on prey capture in toads by Jorg-Peter Ewert. The best stimulus for snapping at prey by a toad is a long thin dark bar, moving across its retina like a worm. If such a bar has a side projection added to it, so that it becomes L-shaped, the snapping response of the toad tends to be inhibited: the worm has become an 'anti-worm'. Ganglion cells

Figure 3.8. Binocular vision of the toad *Xenopus* enables it to assess distance and so capture prey effectively. Fibres from each eye pass to the optic tectum on the far side of the brain, but connections between the optic tecta link cells which receive the same visual input. Thus an object at Z in space stimulates point 1 on the retina of the left eye and point 1' on the right. A cell in the left tectum which receives direct input from point 1' also has an indirect connection via the other tectum to point 1 so laying the foundation for binocular vision. (After M. Keating (1974), *Br. med. Bull.* **30**, 145–51.)

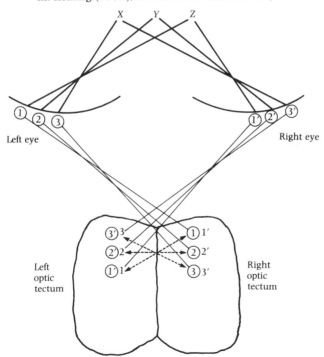

in the retina respond to objects of the right size and speed, but do not discriminate between worms and anti-worms. Instead, this discrimination takes place in the optic tectum: there are cells here which do distinguish between different types of stimuli, though no one of them has all the properties of prey recognition. It seems that the system depends on a network of nerve cells which receives input from the two eyes and from which output goes to the tongue musculature and so gives the response.

Turning now to the hearing of amphibians, we find that the ears of frogs, like their eyes, are tuned to sounds of particular interest to frogs, in this case the calls of their own species. Figure 3.9 shows the detector mechanism which bullfrogs are thought to use. Some receptors in the ear

Figure 3.9. Diagrammatic representation of how bullfrogs detect each other's calls. The call has energy at two different frequencies and cells in the ear respond preferentially to sounds at the two peaks. Intermediate sounds, between 500 and 1000 Hz (cycles per second), suppress response. It is postulated that detection is based on three filters matched to the three frequency bands, high and low frequencies stimulating the detector while medium frequencies lower its output. The detector is an 'and' one: there must be energy at both peaks and there must not be energy in the trough before it will respond. (From R. R. Capranica & G. Rose (1983), in *Neuroethology & Behavioral Physiology*, ed. F. Huber & H. Markl, pp. 136–52, Springer-Verlag, Berlin.)

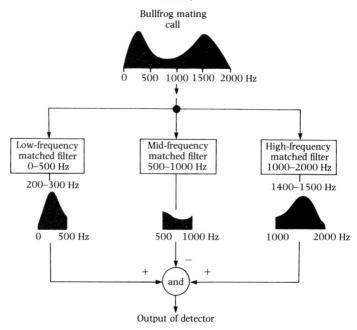

are stimulated especially by tones of just over 1000 Hz (cycles per second), while others respond mainly to sounds at around 300 Hz and their firing is suppressed by frequencies rather higher than this. So, a sound at the lower frequency makes the cell respond but adding another slightly higher pitched sound to the first will decrease its firing. This system matches the animal's hearing perfectly to the two energy peaks in the call of the male bullfrog.

This is a good example of a peripheral filter at work and shows how subtly the sounds animals produce and the senses of their hearers may be matched to each other. Indeed the match may be yet more precise. Cricket frogs produce higher pitched sounds and their calls show dialects, varying in frequency from place to place. But the frequency a male produces in a particular area is matched to that of the female's ear in the same place: 3750 Hz in East Kansas, 3550 in New Jersey, 2900 in South Dakota. Another fine piece of matching between the calling of the male and the response of the female stems from the fact that amphibians are cold blooded. This means that on a cold day the male gray tree frog produces fewer pulses per second than when it is warm, which could obviously lead to problems in mate attraction. However, it transpires from experiments by Carl Gerhardt that the preference of the female varies in an exactly parallel fashion. The only problem comes when she is sitting in a place where the temperature is very different from where he is calling, for then she ignores him: but that particular circumstance only really arises when a prying scientist has changed the temperature of one of them!

3.4 Central and peripheral filters

From these examples it is clear that animals are endowed with the equivalent of filters, which let some stimuli past and so lead to behaviour appropriate to them, while excluding other stimuli which should be ignored or treated differently. Sticklebacks do not attempt to court their rivals, nor do gulls try to roll their chicks into the nest when they run out of it. In some cases the filters appear to be properties of the sense organs, while in others more central mechanisms are involved. We can ask questions here somewhat similar to those we considered in respect of centrally or peripherally organised movement patterns. Why is it then that some of the stimuli to which animals respond are simply those to which their sense organs are tuned, while others are much more complicated combinations of features, the recognition of which requires a great deal of processing by large numbers of nerve cells within the brain itself?

Some filtering obviously goes on in all sense organs as they do not respond to every conceivable stimulus. The fact that we cannot see the ultraviolet patterns on flowers is an example of this. But no one would doubt that, within their range of sensitivity, a great deal of raw information is passed by our eyes and ears back to the brain and that it is there that we decide what to do with it. The problem with more precise sensory filtering, like that involved in many of the examples from amphibia discussed above, is that it can only work where the sense concerned is used in responding to one or a few very specific stimuli. A good example of a sense organ tuned so that it can be used for two different things is in the ear of the cricket *Teleogryllus oceanicus* (Figure 3.10). Like other crickets, these animals show peaks of responding at frequencies matching those in their species song. But they also have units which are tuned to much higher frequencies and probably enable the crickets to detect the ultrasonic cries of a bat which preys upon them. Their ears (which incidentally they keep

Figure 3.10. Threshold plots to show the sensitivity of the ears of three different species of cricket. The lower the points, the more sensitive is the animal at the frequency shown along the bottom (1 kHz = 1000 cycles per second), as the scale on the left is one of loudness (sound pressure level in decibels). *Gryllus campestris* (□) is most sensitive at 4 and 14 kHz, and these are the main frequencies in the song of that species; the peak sensitivity of *Teleogryllus commodus* (■) also matches the energy peak of its song (3.8 kHz). In the third species, *T. oceanicus* (●), the low thresholds at 4–5 kHz and at 20 kHz match the ear to the species song, but there are also sensitive units at higher frequencies (30–40 kHz) and these may aid in the detection of predatory bats. (After M. Hutchings & B. Lewis (1981), *J. comp. Physiol.* **143**, 129–34.)

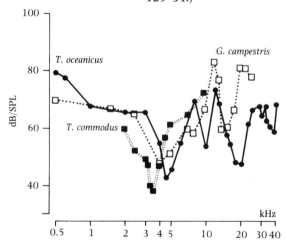

in their elbows!) seem therefore well adapted for detecting both mates and predators.

Despite this example, it is hard to see just how a sense organ could be wired up to allow an animal to distinguish between many different sorts of stimuli. Precise peripheral filters are therefore most likely to be found in animals with few interests in life. The ears of frogs are probably largely concerned with hearing the rather stereotyped croaks of other members of their own species and have little role in detecting food or predators. Their eyes are largely concerned with the distinction between things to eat and things to run away from, though they certainly also have a role in mating once a female has been attracted to close range. However, the information they provide on mates is not very precise: male frogs and toads are notoriously unselective about the objects which they will attempt to clasp during the mating season!

The uses to which herring gulls put their sense organs are much more varied and a great deal of the responsiveness that they show must, accordingly, be centrally determined. Although gulls do respond best to large eggs with many small spots on them, and this could easily be because such eggs stimulate their eyes most, their preference for other aspects is not so easily explained. The favourite colour of egg for the egg-rolling response is green, yet the feeding gull prefers objects which are brown and the gull chick, when pecking for food from its parent's beak, pecks the red spot on the beak and prefers red or blue to other colours. Preference for different colours under different circumstances is unlikely to be accounted for at the level of the sense organs, although many sense organs do have motor nerves going to them from the brain which are probably involved in adjusting sensitivity in various ways and could even swing different types of filter into action.

In birds and in mammals the retina is much less complicated than it is in lower vertebrates, but the visual pathways in the brain are more extensive. Fibres from the retina pass to the optic tectum, known as the superior colliculus in mammals, and this seems to be involved in locating objects in space, as it is in toads. However, there are also connections from the eyes up to the visual cortex of the brain, and here a great variety of different types of cells are found. Many of them respond to edges passing across the retina in particular orientations, and these cells occur in ordered columns, where those near each other have similar preferences. Other cells are binocular, firing to stimuli at equivalent points on the two eyes. Yet other cells have been found with much more complicated and specific preferences: for example, some cortical cells in starlings fire only to a particular species-specific call, and some in monkeys will recognise a face.

While a male toad may confuse a wide variety of objects of about the right size and shape with potential mates, this is not usually a problem with gulls, rats or humans. Indeed these animals may be able to distinguish their mates from all other members of the species using subtle visual, auditory or olfactory cues. Gulls, rats and people are also omnivores and recognise all sorts of foods while rejecting many objects which look similar but are in fact inedible. The distinction between predators and similar species that are harmless can also be slight, yet it creates few problems: many small mammals of the African plains will scurry away at the sight of an eagle, yet feed unperturbed as a vulture passes over. In cases such as these, where many different stimuli have to be assessed and responded to appropriately, the recognition processes involved must depend on the computer power of the brain rather than on simple filters in the sense organs themselves.

4

Motivation

When confronted with an appropriate stimulus, an animal does not always show the same response. As we saw in Chapter 2, even the simplest of reflexes fatigues and must then be allowed a period of recovery before it can be elicited again. However, few of the changes in responsiveness shown by animals can be put down to fatigue. A male stickleback on his own in a tank and shown a model female for a brief period every day may sometimes court it with great vigour, yet on other occasions he ignores it. Anyone who studies animal behaviour soon comes to realise that it varies enormously from time to time, and this is frustrating if one hopes that animals will always behave in the same way. But, on the other hand, discovering just what it is that leads them to behave differently from one time to another presents some very interesting problems which are well worth studying in their own right. These are the problems of motivation or, in other words, the mechanisms leading animals to do what they do when they do it.

What motivates an animal from within is obviously a much more difficult subject to come to grips with than the stimuli from the outside world which affect its behaviour, and it can be studied in two very different ways. One is to look inside and see what is happening there, for example by stimulating nerves or recording from them. The hope here is to find the pathways between the sense organs and the muscles, and the physiological factors which affect them and so influence the behaviour. The other approach, more conventional amongst ethologists, is to treat the animal as a 'black box', changing the inputs to it in various different ways, for example by depriving it of food or introducing a rival to it, and seeing how that alters behaviour, rather than looking inside to see what mechanisms are involved. Just as one need not understand how a computer works to program it and understand the output it provides, a lot can be found out about motivation without any knowledge of the hardware responsible for it inside the animal. This chapter will illustrate some of the approaches that have been used to study motivation in this way.

51

In thinking about motivation, it is worth considering two rather different aspects separately. One of these is how the tendency to behave in a particular way fluctuates over time, and this involves looking at different types of behaviour, and the factors which affect them, without much concern for other activities. As an example of this, one might study how eating is patterned in time and how it is affected by periods of food deprivation. The second approach is to examine how the animal decides which of the many behaviour patterns at its disposal it will perform. In this case one might look to see whether a rat which is hungry and also thirsty decides to eat or drink first. We will consider these two approaches in turn, starting with a simple model which was aimed at trying to account for changes in the tendency of animals to behave in one particular way.

4.1 A model of motivation

We all experience hunger at some times and not at others and so we realise that the urge to eat is not constant. The sight of a delicious meal is on some occasions mouth-watering and on others of no interest at all. Once again we owe to Konrad Lorenz a clear theory of how such changes might come about which helps us to think about the problem. His theory was put forward in the form of the model shown in Figure 4.1, the word 'model' in this sense meaning a simple scheme which is proposed to work in a way similar to the real system. Lorenz's theory is known either pretentiously as his 'psychohydraulic' model or, more prosaically, as 'Lorenz's water closet'.

This model is not something one would look for inside an animal! It is what is called an 'as if' model: the animal may behave as if it had such a system for organising its behaviour within it. The theory helps one to think about the problem and design experiments to see how the system really works rather than proposing specific mechanisms. Lorenz supposed that different actions depended for their appearance on a supply of 'action-specific energy' which accumulated with time since the animal last behaved that way and was used up as it performed the act. He visualised this as water accumulating in a tank out of which it could only escape through a valve at the bottom. The valve was, however, a rather strange one. It could be opened either by the water pushing within, or by a string attached to a scale-pan pulling from outside. To Lorenz, weights on this scale-pan were the equivalent of stimuli leading to the behaviour: the more appropriate the stimulation, the heavier the weights and the more likely the valve was to be opened.

Lorenz's model has some interesting properties which can be compared with real behaviour. First, the longer since the behaviour was last performed the more action-specific energy will have accumulated and the more likely it is that the behaviour will appear. Secondly, the model suggests that the accumulation of action-specific energy will lead the behaviour to occur even if the stimuli present are slight (the weights on the scale-pan are very light). Ultimately, Lorenz argued, behaviour of which an animal has long been deprived will appear as a 'vacuum activity' with no stimulus present at all.

Figure 4.1. Lorenz's 'psychohydraulic' model of motivation. Action-specific energy is represented by water, which accumulates progressively in a reservoir when the behaviour concerned is not being expressed. The behaviour pattern occurs when water passes out of the reservoir into the trough beneath, higher threshold aspects of the behaviour (numbered 4, 5 and 6) only being shown when a lot of water is passing into the trough. The valve is so arranged that it is opened by the combined effect of water in the reservoir (action-specific energy) and of weights on a scale pan, which represent the adequacy of external stimulation. (After K. Z. Lorenz (1950), *Symp. Soc. exp. Biol.* 4, 221–68.)

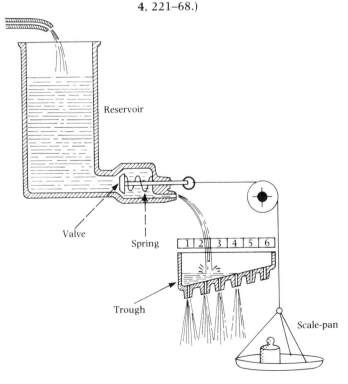

The idea of vacuum activities came at one extreme: when there was an excess of action-specific energy. At the other was exhaustion which occurred, according to Lorenz, when an activity had been stimulated so often that the animal had run out of this energy. Then, there is a final idea incorporated in the model. The way the valve works suggests a particular relationship between internal factors and external stimuli: provided that there is some action-specific energy, the push of this and the pull of external stimuli will add up to give rise to the behaviour, rather than being multiplied together or related in some more complicated way.

How does real behaviour match up to these four suggestions that Lorenz made? Let us consider each in turn.

4.1.1 Accumulation of energy

The nervous systems of animals have no stores of energy within them as Lorenz proposed, and this is obviously not a realistic aspect of the model. Nevertheless, animals might behave as if they did and, if so, we would expect them to become more likely to perform a particular action with the passage of time since they last did so. Do they do this?

There is no doubt that certain aspects of behaviour do become more likely the longer the gap since they last appeared (see Figure 4.2). Feeding and drinking are the most obvious of examples, as we all know from our personal experience. There are very good reasons for this: nutrients are used up by the body and water is lost constantly from it, so both must be replenished and the need to do so will rise with the interval since the last meal or the last drink. But many of the other activities which animals perform are not concerned with regulating aspects of physiology, so there is no reason to think in advance that they might become more likely with time, and they are quite often found not to do so.

Many behaviour patterns, such as singing in birds, mating in fish or exploration in rodents, could be used to illustrate this point. However a particularly appropriate one to take is aggression. Although he did not mention his psychohydraulic model in his book *On Aggression*, Konrad Lorenz obviously had it in mind. He suggested that aggression is an innate drive which rises with time and must somehow be expended. His view of innate behaviour is that it is inevitable (a point taken up in the next chapter) and the only alternative is to sidetrack it into harmless channels: for human aggression he suggests that sport may play such a role, helping us to get rid of otherwise destructive urges.

Leaving aside the issue of whether sportsmen are less aggressive than other people as a result of their activities, there are a great many objections to these views. Some of them concern whether the urge to behave aggressively does accumulate in an inevitable fashion. One of the few cases where it seems to do so is where an animal is isolated, for example a mouse placed in a cage on its own. The longer it has been in isolation the more it will fight another mouse which is put in with it. But this could simply be that, when in company, it habituates to the other animals with it and it needs some days on its own to recover its tendency to fight them. In fact some fish fight less after isolation and need stimulation from others over a period before the urge to fight returns (Figure 4.3): in this case contact with others probably leads to the gradual build up of a hormone which makes aggression more likely.

Many factors affect aggressive behaviour, such as hormones and shortage of food on the inside, and presence of rivals and contested resources on the outside. Furthermore, aggression is itself a complex of different actions which may be quite differently caused, though they are superficially similar. It is not an easy matter to decide the extent to which

Figure 4.2. Traces from a pen recorder to show the feeding behaviour of 12 rats over a period of about 2 hours. The animals tend not to nibble at food the whole time, but pattern their feeding into meals, taking one of these at fairly regular intervals. Thus the longer it is since its last meal, the more likely it is that a rat will eat. (After D. W. Thomas & J. Mayer (1968), *J. comp. physiol. Psychol.* 66, 642–53.)

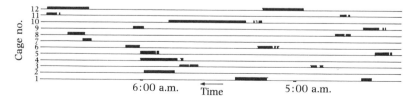

a hawk killing a mouse, a cornered subordinate rat defending itself against a dominant one, two fighting cocks locked in a struggle, and a mother duck defending her chicks from the unwelcome attentions of a gull, are actually motivated in a similar way. Aggression serves many different functions within one species (Figure 4.4), and its uses and the situations in which it appears may vary between species, so that the mechanisms underlying it are also likely to vary. Evolution matches behaviour marvellously well to an animal's particular environment and way of life. So it should be no surprise that the organisation of aggression varies a lot between species, just as feeding differs between the lion that kills and eats every few days and the wildebeest that must crop its plant food for most of its waking hours. Likewise, there is no reason why the urge to behave aggressively should mount with time like hunger or thirst, and it is rather surprising that Konrad Lorenz should have thought this, given that he is a biologist with great respect for the power of evolution to adapt behaviour to an animal's exact requirements.

Figure 4.3. Males of the cichlid fish *Haplochromis burtoni* seldom attack small fish kept in their tank but, if a model of another large male is put in with them for 10 days (black bar), their attack rate slowly rises, then falls back again gradually after the model is removed. Thus they do not attack most as soon as the model is presented, as would be expected had they accumulated action-specific energy in its absence, but instead need to be stimulated by a period of its presence before their attack rate rises. (From W. Heiligenberg & U. Kramer (1972), *J. comp. Physiol.* 77, 332–40.)

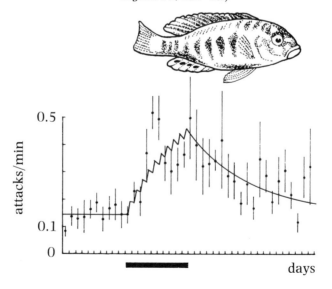

Figure 4.4. A cat cornered by a dog shows defensive behaviour, fighting back as best it can (*a*). Cats will also attack mice they encounter, here pouncing and showing other predatory actions as well as using their teeth and claws (*b*). In territorial disputes cats may fight intruders of their own species (*c*). Many of the actions involved in these different situations are similar. Although all of them involve attempting to inflict damage on other individuals, they probably have very different causes, so that it is not useful to think of them all simply as examples of 'aggression'.

(*a*)

(*b*)

(*c*)

A final aspect of the accumulating energy idea concerns the extent to which an animal needs to perform the action as opposed to needing to achieve its results. Must a hungry animal make the appropriate number of movements, as Lorenz's model would propose, or does it just need to have the right amount of food inside it? Does an aggressive animal have to perform a certain amount of fighting, or will it cease to be aggressive when its rival is repelled? In Chapter 2, we saw that animals tend to regulate their behaviour as they go along in the light of its consequences: a process called feedback. The dog does not rush at the place where the rabbit used to be but follows an arc so as to close upon it, changing its course as required. Unless its muscles are fatigued, it does not stop running before it reaches the rabbit, nor does it carry on after it has caught it. This is quite different from the action-specific energy idea, which proposed that the animal should perform the amount of behaviour for which it has accumulated energy regardless of its consequences. This would suggest that an eating animal should chew and swallow a set number of times even if an experimenter had surreptitiously raised the glucose in its blood or filled its stomach with food. In fact, animals do not generally do that: unlike the egg-rolling goose, they respond to feedback from their actions. They drink until receptors tell them that they have taken in enough water and they eat until they have had enough food, rather than showing a fixed amount of behaviour depending on the time since they last ate or drank.

4.1.2 Vacuum activities

There are few examples of the idea that animals deprived of an opportunity to perform an action might eventually show it even in the absence of all stimuli. Lorenz himself described a pet starling that would flutter up to the ceiling to catch a non-existent fly, but it is hard to be certain that there was nothing there to which the bird was reacting. There is no doubt that behaviour can sometimes show stimulus generalisation, so that animals deprived of the usual stimulus will show the behaviour to a much less adequate one. In zoos, they will often mate with the wrong species when their own is not available: lions and tigers, for example, will mate to produce 'ligers'. As we all know, the hungrier one is the more one is prepared to eat less tasty food (indeed, if very hungry, I suspect even I might eat rice pudding!). Thus, generalising to less adequate stimuli is a reality, but there is not so much evidence for behaviour being shown in the total absence of any appropriate stimulus, as the vacuum activity idea suggests. Perhaps the best is an account by the ornithologist David Lack of the behaviour of a European robin he had just finished testing with a stuffed bird of its own species, a potent releaser of aggression for these birds. As he removed the specimen and walked off, he chanced to look back and saw the robin fluttering around, singing and delivering pecks to the position in mid-air where its apparent rival had been (Figure 4.5). Here the appropriate stimulus had clearly been removed, but the animal was certainly not suffering from deprivation of the opportunity to behave aggressively!

4.1.3 Exhaustion

In Chapter 2, I discussed how reflexes might cease to be performed through muscular fatigue, through sensory adaptation, or through some process of neuronal exhaustion which took time to recover. Could the last of these simply be like running out of action-specific energy? Several detailed studies cast some doubt on such a simple view. A good one with which to start is a study of water boatmen by the Dutch ethologist H. Wolda (Figure 4.6).

Water boatmen are insects which lie suspended upside-down just beneath the surface of a pond waiting for slight disturbances which may indicate the presence of prey. If the surface near one of these animals is pricked with a pin, the insect will turn towards the ripple, but after many such pricks it ceases to respond and takes a long time to recover. Even 24 hours later, a series of tests will not give as many positive responses as in a fresh animal (Figure 4.6b). Nevertheless, there are a number of reasons for thinking that this is not action-specific exhaustion. One is that an animal which has become fatigued by a series of 300 pricks on the water surface to the left of its head will start to respond strongly once more when

Figure 4.5. A European robin attacks the spot in mid-air where previously a stuffed rival had been placed.

the stimulation is moved to the equivalent position on the right side. Thus the exhaustion is specific to the location of the response and so does not apply to all responding. Another point is that recovery only occurs if stimulation ceases (Figure 4.6c). An exhausted animal will start to respond

Figure 4.6. (a) The simple laboratory set-up used to study the response of the water boatman (*Notonecta*) to pin pricks on the surface of the water near it. (b) The number of responses in eight series of 25 trials, in a fresh animal (1), one allowed to recover for 24 hours since previous testing (2), and one given only 5 minutes to recover (3). The animal responds much less towards the end of the 400 trials, even if fresh, but shows some recovery when rested. However, even after 24 hours, its recovery is not complete and it becomes exhausted more rapidly than it did the previous day. (c) Recovery will not occur if the stimulus continues, even if the animal does not respond to it. (After H. Wolda (1961), *Arch. neerl. Zool.* **14**, 61–89.)

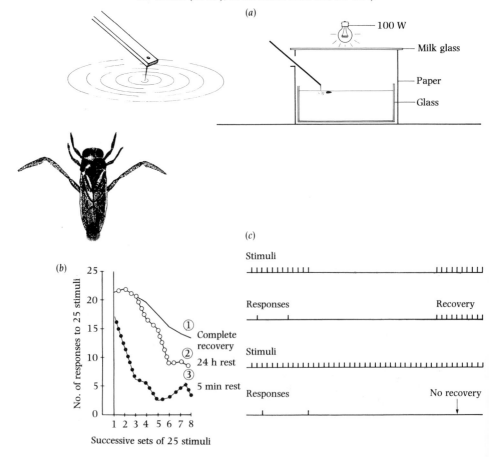

again after a few minutes if the water surface suddenly starts to be pricked again. But, if the stimulation has continued during these few minutes, recovery will not occur. In other words, it is not that a period without the action is required for recovery, but that the animal needs to be free of the stimulus for some time. This is stimulus-specific waning of response or, in other words, habituation, and it is a very common phenomenon. We all know how one ceases to notice sounds, like that of a ventilator fan running continuously, but one jumps to attention when it stops or its tone changes.

Ceasing to respond need not therefore indicate that the action is exhausted. Furthermore, in some other cases, behaviour which is repeatedly elicited shows signs of becoming easier, rather than more difficult, to produce: this is a process of sensitisation rather than exhaustion. The mating behaviour of male rats is an example which shows both processes (Figure 4.7). If a male that has not mated for some time is placed with a receptive female, he will perform a series of mounts at the end of which ejaculation occurs; he will then groom for a while before a second series, and he may achieve three or four ejaculations in the course of an hour. In keeping with the idea that he is becoming exhausted is the fact that his refractory period, the time off he takes between mount series, increases from one copulation to the next. However, the number of mounts he requires to reach ejaculation actually falls as the encounter proceeds: to begin with he needs a lot of stimulation; later on he needs much less. As far as this aspect of his behaviour is concerned, he is becoming sensitised rather than exhausted.

Examples such as these suggest that the changes in behaviour which

Figure 4.7. The mating behaviour of a male rat placed with a female until he loses interest in copulating. On average he needs over 10 intromissions to achieve ejaculation the first time, and then starts mating again about 5 minutes later. After a few ejaculations only half as many intromissions are required, but it takes him twice as long to recover. (Data from F. A. Beach & L. Jordan (1956), *Q. J. exp. Psychol.* **8**, 121–33.)

Mating episode	Number of intromissions needed	Time to start mating again (seconds)
1st	10.6	324
2nd	6.0	395
3rd	5.7	468
4th	5.1	495
5th	5.1	597
6th	4.1	818

animals show when repeatedly stimulated are far from simple and certainly cannot be thought of in terms of them running out of action-specific energy. Sometimes they become more prone to show an activity rather than less; exhaustion of some aspect of behaviour may be rapid, of others slow; in some, recovery depends on not showing the action, in others, removal of the stimulus is required. Even in the male rat which has copulated to exhaustion, it is not simply that his internal state has become incapable, as if he had run out of action-specific energy. Give him a fresh female and he will start mounting again!

The conclusion then is that Lorenz's model only shows a very superficial similarity to the behaviour of animals as far as exhaustion is concerned. Their changes in responsiveness with repeated stimulation are far more complicated than the simple running out of energy that the model shows.

4.1.4 *Internal and external factors*

The last important point raised by Lorenz's model is that of the relationship between internal and external factors. The model suggests that the two add up to give behaviour, as both weights on the scale-pan and water in the tank can force the valve open, except that the tank must contain a minimum of water if the behaviour is to appear at all.

Just how internal and external factors relate to give behaviour is still a matter of contention and it may well be that some actions involve different relationships from others. It is certainly the case that behaviour patterns vary in the relative importance of internal and external factors in bringing them about. Feeding and drinking obviously need both: a rat will only drink if it is thirsty and if water is present; it will only eat if it is both hungry and presented with food. But consider two other examples: an exploring mouse and a singing bird (Figure 4.8). Mice kept in small cages explore rather little, but if they are suddenly presented with a new area into which they can wander they will start to move around it, sniffing into corners and rearing up onto walls. The larger the area, the longer they will spend looking around it. Thus exploration is quiescent for months on end, and then suddenly it emerges when a new environment presents itself. It is therefore a behaviour pattern largely driven from the outside rather than fired from within. The song of birds is just the opposite of this. Most male songbirds sing, provided that they are in breeding condition, and a critical factor here is usually the male sex hormone testosterone. They will sing as long as they have enough of it in the bloodstream. The outside world is less important, but it does have some influence. Males may only sing on certain song perches, and they may sing more when it is warm and sunny, but the urge to sing is mainly from within them. Thus song and exploration represent two behaviour patterns at the ends of a spectrum, one mainly dependent on inside stimuli and the other on outside ones. Most other behaviour patterns, like feeding, grooming or sexual behaviour, lie somewhere in between these extremes and rely on both types of influence to a varying extent.

Regardless of which are the more important, just how do internal and external factors interact to give behaviour? Do they add together as Lorenz's model suggests? Once again it seems possible that the relationship is quite different for different types of behaviour. Addition is certainly a possibility, though it is an unlikely one because it suggests that behaviour may appear with no outside stimulus as long as the internal factors are high and, as discussed above, such vacuum activities are certainly not common. It also suggests that a high level of external stimulation might lead an animal to show the activity even if internal factors are absent. There is even less evidence for such an effect. Nevertheless it is extremely difficult to tell how internal and external factors do relate to one another. The problem is to decide 'how much' of each there is. The courtship of guppies will illustrate this point (Figure 4.9).

Male guppies court large females more vigorously than they do small ones. The male also indicates his internal state, as do many fish, by changing his colour patterns to become more brilliant when he is more highly motivated. Thus the colouration of the male gives a measure of his internal state and the size of the female is an indication of how good a stimulus she is. Plotting one measurement against the other shows that they are clearly related. Males which are not well motivated will only

Figure 4.8. Two behaviour patterns with very different motivational bases. The stimulus for a mouse to explore lies largely in the outside world, while internal state is the major factor leading to song in birds.

display to females which are very large. On the other hand, males which are exceptionally brightly coloured will even display to very small partners.

This shows that internal and external factors interact to give rise to behaviour, but it does not tell us how. The size of a female could be assessed by her length, the area she projects onto the retina of the male, her weight, or many other measures. Assessing the colour patterns of males is even more difficult. Clearly some of them are more likely to lead to display than are others, so they can be put in an order, but there is no way that their distance apart can be accurately determined to give a scale of motivation. Thus, even after careful study showing how internal and external factors relate to one another in guppy courtship, it is not possible to say whether they add together, multiply, or have some more complex relationship.

Figure 4.9. Three displays shown by male guppies are posturing (P), intention sigmoid (IS) and full sigmoid (S). The graph shows the size of female needed to elicit each of these from males with various different markings. These provide a guide to the male's internal state, those on the right being more highly motivated than those on the left. (After G. P. Baerends, R. Brouwer & H. T. Waterbolk (1955), *Behaviour* **8**, 249–334.)

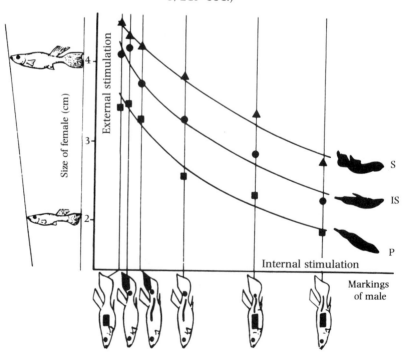

This consideration of Lorenz's model of motivation shows just how far it has proved inadequate as a result of the many years of research which have elapsed since it was originally proposed. It was at that time a reasonable hypothesis, and has proved since to be a fruitful one. But most of its features have been found to be only crude approximations to the way that behaviour is really motivated. Two flaws stand out perhaps more than others. First, it has no role for feedback from the consequences of behaviour, an influence now realised to be extremely important for most behaviour. Secondly, it presents a picture of motivation which is the same for all behaviour patterns. Later research has shown just how different many of these are from one another and how dangerous it is to assume that the organisation of one is similar to that of another. Song, feeding, grooming, mating, drinking, fighting and exploration all depend on their own particular internal and external causes and, furthermore, these may well differ from one species to another.

4.2 Deciding what to do

Animals generally only do one thing at a time, yet they often have the need to perform several. For example, a caged zebra finch waking up after a 12 hour night must have a long list of priorities. These birds normally eat and drink about every half-hour, sing and fly around their cage for a period

Figure 4.10. The short-term cycle of behaviour shown by a male zebra finch. The illustration is based on observation of 12 cycles shown by a single bird synchronised at 30 minutes long by a cycle of changing light intensity. In each cycle the bird moves from resting to an active period during which it feeds, through a peak of singing to grooming, and then back to resting once more. (After P. J. B. Slater & A. M. Wood (1977), *Anim. Behav.* **25**, 736–46.)

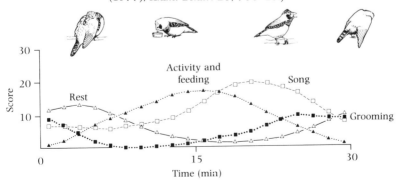

after this, and groom and rest till the next meal is due (Figure 4.10). Most of their grooming is brief, but every two hours or so they show a long bout which deals with all the areas of their body. At night they groom a little and rest a lot, but none of their other activities is performed during the hours of darkness. Not surprisingly, then, they are rather busy in the first half-hour of the morning. Typically, they will stretch and ruffle their feathers when the lights come on, and then move over to their food for a long meal, perhaps followed by a drink. Some flying around and singing will follow this before they settle down for the first long grooming bout of the day.

Despite their need to do several things, these birds show a well-organised sequence of behaviour. They do not rush around doing a little bit of one thing and a little bit of another, nor do they try to sing and groom at the same time. The sequence shown by different birds is similar and suggests that they all have the same priorities. The question posed in this section is how they decide between them, and it is a difficult question to which there is as yet no definite answer. What is clear is that the animal must make decisions and that to do so it must have some means of weighing against each other the internal and external factors relevant to different activities.

The realisation that some sort of weighing up of options must go on within animals was one reason why psychologists postulated the existence of internal 'drives' which, like Lorenz's action-specific energy, were thought of as powering behaviour and were also seen as being measured against each other, the strongest being the one that was expressed. A similar idea was developed by early ethologists, though they tended to use the word 'instinct' rather than drive. Animals were thought of as possessing a small number of instincts, such as those for feeding, aggression and reproduction, and it was these that competed with each other. The instincts themselves controlled a number of behaviour patterns: for example, the reproductive instinct led to various activities concerned with nest building, courtship and parental behaviour. Tinbergen developed a model suggesting that each instinct might be organised in a hierarchy (Figure 4.11). Energy flowed downwards within this from the controlling centre at the top to centres responsible for groups of related activities and then further down to those concerned with individual actions. At each level the appropriate releasers had to be present, these acting to open gates through which the energy could flow to the level beneath and so on to give behaviour.

This is a ingenious idea, and unlike the 'as if' model Lorenz proposed, was based by Tinbergen on the ideas of physiologists; he hoped it might have some neurophysiological reality, with various centres in the brain devoted to different instincts. Unfortunately it turns out not to be that simple. As mentioned before, the nervous system does not store and use up energy as these models suggest. Nor, unfortunately, is it neatly compartmentalised into centres and pathways with clear and distinct behavioural functions. Furthermore, though to some extent behavioural systems, like those controlling feeding, drinking or sexual behaviour, can be thought of as distinct, the factors affecting them overlap and they often influence each other. For example, the hormone oestrogen is an important internal factor making female rats receptive to the male, it also makes them

Figure 4.11. The essence of Tinbergen's hierarchical model was that a centre in the brain controlled each 'major instinct'. Energy from this flowed down to a series of lower centres when a gate was opened by the presence of appropriate releasers. In his original scheme, Tinbergen saw the hierarchy being extended right down to the level of muscle units. Though the hierarchies of different instincts were separate, lack of releasers for one could lead energy from it to 'spark-over' to another and so cause a displacement activity to occur. Thus, as in the example shown here, if reproductive activities were thwarted, grooming might appear.

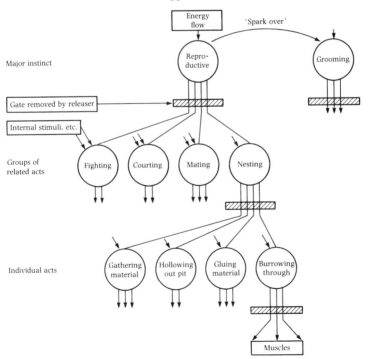

very active so that they are more likely to come across a male, and it makes them less interested in food so that, when receptive, they spend less time eating and more looking for mates. This hormone therefore influences systems concerned with feeding, with activity and with sexual behaviour.

Rather than thinking of animals as having a number of discrete drives, it is usual nowadays to think of them as having overlapping behavioural systems, each of which is responsible for a group of related behaviour patterns (Figure 4.12). Actions with similar functions, like all the different movements involved in grooming or those concerned with nest building, tend to occur in close association, so it is reasonable to think of them as depending on a common system or, to use a more precise expression, as sharing causal factors. A causal factor might be a hormone or a particular level of food deprivation within the animal, or it might be a relevant outside stimulus, like the presence of irritation on its coat or of a member of the opposite sex. On this view of behaviour, whether or not a particular activity is shown depends on the combined level of its causal factors relative to those of all other possible actions.

For several reasons this is a more satisfying way of describing the motivation of behaviour than simply to consider it to be the result of a series

Figure 4.12. Current views of motivation do not see animals as endowed with a series of separate 'drives' or 'instincts'. Instead different aspects of behaviour are thought of as influenced to varying degrees by numerous internal and external causal factors. Here some possible stimulating effects of five different causal factors on three three-spined stickleback patterns are shown.

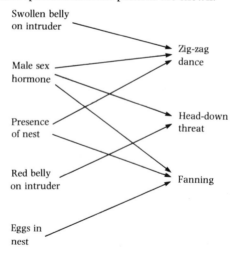

of drives. One of these is that the drive idea is only useful if all aspects of the behaviour can be thought of as affected by the drive in the same way. Careful studies of motivation run counter to this. Most obviously, a causal factor may raise some aspects of behaviour and lower others, whereas if it was enhancing drive, all aspects of the behaviour should be increased. Thus a very hungry animal may chew less and swallow more, and a bird building a nest actively may fly to and fro with material at a great rate, but spend less time than usual carefully selecting and gathering material or in weaving it into the structure.

Another reason why viewing behaviour as dependent on a variety of internal and external causal factors is useful is that it helps to account for why behaviour patterns occur in association with one another to varying degrees. A grooming animal will wash its face, scratch its flank and shake its body, and these very different actions tend to occur close together rather than in association with feeding or mating movements. The most likely reason for this is that grooming actions share causal factors with each other, such as activity in a particular part of the brain or irritation of the coat. Another reason why different behaviour patterns occur in association may be, quite simply, that one causes another. The fact that birds wipe their beaks after drinking almost certainly comes into this category: drinking makes the beak wet and so produces a stimulus which is only removed by wiping. Thus, as well as behaviour patterns sharing causal factors and thereby appearing close together, they can be linked because one acts as a causal factor for another.

The way in which behaviour patterns may compete for expression has been more extensively studied in the case of feeding and drinking, because these are two acts the causal factors for which can be manipulated by depriving animals of food and water for different periods of time. An animal deprived of water tends not to eat as much (doubtless dryness in the mouth makes it difficult to chew and swallow), so the two actions are not totally independent of each other. But just how hungry and thirsty a deprived animal is can be established by seeing how much it eats and drinks when given food and water. We can then say it was 10 grams hungry and 1 gram thirsty if these are the amounts it consumes.

The most detailed experiments on the relationship between feeding and drinking have been those on Barbary doves (known as ring doves in North America) by David McFarland. His research is a good example of how the use of techniques developed by psychologists has helped to illuminate ethological problems. The doves he studied were placed in the 'Skinner boxes' that psychologists often use to study learning, in which they could peck one key to receive water and another to receive food. He found that a dove deprived of both food and water would alternate eating and drinking until it had satisfied both requirements. Its food and water needs, and how these changed as it ate and drank, could be neatly illustrated in the form of 'state space' plots such as those in Figure 4.13. If the bird was more

Figure 4.13. The way in which a deprived Barbary dove replenishes its reserves when placed in a Skinner box where it can peck keys to earn food and water. (*a*) The food and water debt of the bird can be represented by a point in a 'state space'. Here three such points are shown for birds deprived of food alone, food and water, and water alone. Birds eat less when deprived of water, so the point for the last situation shows a small food debt as well. (*b*) A bird deprived of both food and water eats and drinks alternately, so tracing a trajectory in its state space which takes it down to the point where it has cancelled both debts. (*c*) The boundary drawn to separate areas where the animal starts to feed (food dominant) from those where it starts to drink (water dominant). The animal tends to feed until it is well into the water dominant area and vice versa, rather than dithering between the two, as shown in (*d*). (Adapted in part from D. J. McFarland (1971), *Feedback Mechanisms in Animal Behaviour*, Academic Press, London.)

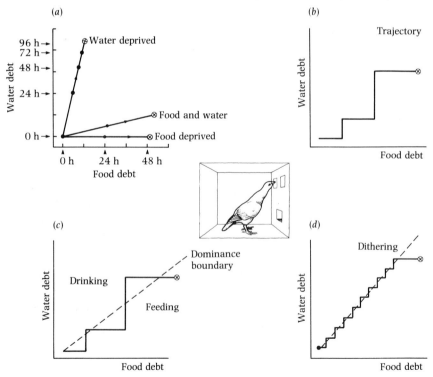

hungry than thirsty it would start by eating (as in Figure 4.13*b*), and the amount that it ate could be traced by a line on the graph. After eating for a little it would then become more thirsty than hungry and would switch to drinking, the line on the graph moving so that the water deficit was shown as decreasing, while that for food remained the same. Finally, after several drinks and meals, the line would have zig-zagged down to zero on both axes, indicating that the animal was satiated.

On this graph we can assume that every time the animal switches from one action to the other it is moving to that which has greater causal factors. We can draw a diagonal line which gives a rough impression of the boundary between feeding having priority and drinking doing so (Figure 4.13*c*). All switches in the drinking segment are from feeding to drinking, and those in the eating one are in the other direction. A feature which may seem curious here is that the animal goes well over the line in each direction before it switches rather than doing so immediately. If it did the latter, it would end up 'dithering': after an initial large meal or drink to take it over the boundary line, it would oscillate rapidly from one activity to the other, taking tiny amounts of each in turn until eventually both needs were satisfied (Figure 4.13*d*). This would obviously be a badly organised way of doing things, but what makes sure that it does not happen? Several factors are probably involved, but one is likely to be especially important. This is that an animal that has just taken an item of food is likely to be looking at a food dish rather than at stimuli appropriate to any other action. As the sight of food is a causal factor for feeding, and that of water for drinking, this will mean that the animal is likely to continue with what it is doing rather than switching to another behaviour for which internal factors may well be high, but external ones are absent.

In this simple example, the state space plot expresses the relationship between two activities, feeding and drinking, but a third dimension might be added so that the animal's tendency to groom was also mapped in the hope that one might be able to predict when it would take a break from feeding and drinking to clean itself. Ultimately, one might hope to achieve mapping of all behaviour patterns onto each other in a 'multidimensional state space' but, as well as it being difficult to imagine such a complicated diagram, it is not easy to see what currency they could all be expressed in. There is certainly no such thing as 'grams' of grooming or of tendency to fight. So the problem of how the nervous system weighs up different actions against each other remains as real as ever.

4.3 Displacement activities

The idea that animals sometimes show displacement activities was a final proposal made by ethologists which deserves discussion. Watching a pair of courting birds, the observer might be surprised to see one of them break off to preen briefly and hurriedly before carrying on with its courtship. Similarly, a fish in the middle of threatening a rival may suddenly swim down and start digging in the sand as if switching to nest-building, or a cock might interrupt fighting to take a few pecks of food (Figure 4.14). Such actions seemed irrelevant to those who saw them, and they often shared other features such as incompleteness or a hurried and frantic appearance, so they came to be grouped together and labelled as displacement activities. It was suggested that they arose when the animal was unable to carry on with more relevant behaviour because it was in a conflict as to what to do or was thwarted from achieving its aims. Thus the fighting gull might be in a conflict between attack and escape and so unable to do either; a courting male fish might have his sexual approaches thwarted by an unreceptive female and so, again, be unable to show the behaviour most relevant to the moment. In line with his hierarchical model of motivation, Tinbergen suggested that the thwarted energy from the reproductive instinct 'sparked over' and motivated behaviour down a different channel, such as that concerned with grooming (see Figure 4.11).

Ideas on displacement activities have changed enormously since they were thought to be due to sparking over. An important finding was that the causal factors which affect the behaviour in its normal context also

Figure 4.14. Three classic examples of displacement activities: ground pecking in a fighting cock (*a*); sand digging in a three-spined stickleback confronted by an intruder (*b*); wing preening in a courting mallard drake (*c*). (After N. Tinbergen (1952), *Q. rev. Biol.* **27**, 1–32.)

(*a*)   (*b*)   (*c*)

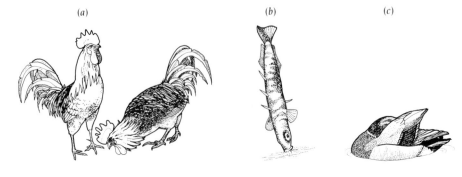

affected it when it appeared as a displacement activity. A courting male stickleback might break off to fan at the nest, a behaviour which serves to aerate his eggs but which is irrelevant when there are no eggs there yet. Nevertheless, in its courtship context as well as when it appears later, the fanning is enhanced by carbon dioxide in the water. In the same way a bird which grooms when in conflict shows more of this grooming when its feathers are wet. Results such as these show that the behaviour patterns labelled as displacement activities are motivated in the usual way: their irrelevance lies only in their context.

The behaviour patterns labelled as displacement activities are undoubtedly a rag-bag of different actions, differently caused, the main thing they have in common being that a watching ethologist thought they were out of place. Conflict or thwarting may certainly be one reason why they arise: the animal, unable to carry out actions which are highest in priority, moves instead to ones which are next in line. This process is called disinhibition, and it is probably an important reason why displacement activities appear. It is perhaps not surprising that very often the displacement activities that have been described are grooming movements, because the animal carries around the stimuli for grooming with it so that, unlike food or water, they

Figure 4.15. A hungry guinea pig placed in a conflict by a rotating card near its food dish stops some distance from its goal and shows various behaviour patterns such as grooming.

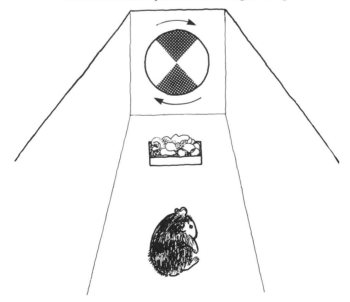

are always there on the outside of its body. Thus, if disinhibition occurs, grooming is very likely to follow. However, another reason why grooming often occurs out of context may be much simpler: the vigorous actions of fighting or courtship that the animal has been showing may lead its fur or feathers to be dishevelled and so actively enhance the stimuli for grooming till it breaks off for a quick preen to get rid of the itch.

In the days when ethologists thought of behaviour as being motivated by instincts or drives, it was these concepts that were thought to be in conflict with each other within the animal, or to be thwarted by a barrier to its behaviour on the outside. Thus conflict between the drives of aggression and of escape, or thwarting of the sex drive, were commonly referred to. However, these ideas were very vague: the drives themselves were hypothetical, so the conflict between them was doubly so. The theory was also virtually untestable: if everything was due either to its own drive or to a conflict between other drives, it was hard to think of any observation or experiment that could disprove it. Thus, as drive theories have fallen from favour, so has the idea of conflict between drives. But this does not mean that conflict or thwarting is not a real phenomenon and an important cause of switches in behaviour. Thwarting can easily be arranged by placing a perspex screen between a hungry animal and its accustomed food dish. Similarly, an approach–avoidance conflict can be set up by placing a frightening object beside a food dish (Figure 4.15). In such circumstances a whole array of behaviour patterns may be disinhibited because the animal is unable to perform those that are its top priority.

4.4 Conclusion

Motivation theories have come a long way since the original ideas put forward by Lorenz and Tinbergen over 30 years ago, yet the topic remains a difficult one. The study of individual systems, like those controlling eating, sexual behaviour or exploration, has advanced mostly in the cases of those in which the internal state of animals can be altered physiologically to discover just what causal factors do operate from within. The influence of hormones on sexual and aggressive behaviour, and of nutrients and ion balances on eating and drinking, have been studied extensively, as have the neural pathways involved in all these actions. Nevertheless, behavioural work aimed at understanding how internal and external factors interact, so that behaviour may be predicted with greater certainty, remains useful and important as a complement to these physiological studies. The interactions between behaviour patterns and the decision-making processes which lead animals to select which acts they will perform are even more complex, and here studying physiology will be of little help. Searching for the rules which govern switches from one behaviour pattern to another is the only option: progress has been made, but the problem is still a daunting one.

5

Development

The problem of how behaviour develops is amongst the most contentious in the whole field of animal behaviour. The last few chapters have illustrated the stress laid by ethologists on words such as 'innate', 'inherited' and 'instinctive' to describe the behaviour patterns they studied. Thus the German word for fixed action pattern, *Erbkoordination*, means, literally translated, inherited coordination. The sensory routes through which these actions are elicited were termed innate releasing mechanisms, and the motivational systems controlling behaviour from within were often referred to as instincts. All these terms implied great rigidity in behaviour and its development, with no scope for the environment to have any influence, as if the action were utterly inevitable and its whole form and development simply a readout of genetic instructions.

On a view such as this, development was obviously rather an uninteresting process, the appropriate behaviour springing fully formed from the animal at the time that it was required without the need for complicated developmental processes involving interactions with the environment. But it was not a view that went uncontested and, in particular, it contrasted with that of the 'Behaviorist' school of psychology in America, which stressed the prime importance of learning in the development of behaviour. For a long time the main issue in behaviour development was this controversy between learning and instinct, so we must start here by considering these opposing views before discussing more recent work on exactly how behaviour does develop.

5.1 Psychologists and learning

The major figure in the Behaviorist school of psychology was J. B. Watson. He followed the English philosopher John Locke in believing that the innate mind was a '*tabula rasa*', a blank slate on which subsequent experience was written. Watson studied learning in animals such as rats and was impressed by the flexibility of their behaviour and by how learning could

76

lead them to behave in an adaptive manner. He believed that they built up connections in their brains as a result of perceiving associations in the outside world, so they would associate a particular place with food and run through a maze to get there. His views were partly stimulated by the work of Ivan Pavlov, the Russian physiologist, who had studied conditioned reflexes in dogs. Pavlov showed how a new stimulus could be made to elicit a reflex as a result of the animal building up an association. In his most famous experiment (Figure 5.1), a dog was presented with food, a stimulus which normally makes dogs produce saliva in readiness to eat, and at the same time a bell was rung. This normally has no effect on salivation. After a number of such presentations, the bell was suddenly sounded without any food being produced. The dog salivated. The animal had formed an association between the bell and the food so that both were now capable of producing the response. This sort of training is referred to as conditioning and the result, in this case salivation in response to a bell, as a conditioned reflex (Figure 5.2).

The main stress in the theories of Pavlov and Watson was not on rewards: the dog did not need to eat the food to form the connection. Later ideas have suggested that a great deal of learning does rely on reward, the most notable exponent of this view being B. F. Skinner. Like Watson before him, Skinner believes that the great majority of behaviour, with the exception of simple reflexes, results from experience. But he stresses the way in which the responses of animals come to be linked to particular

Figure 5.1. The dogs studied by Pavlov were held in harnesses with one of their salivary glands connected through a tube to a collecting bottle. He was able to present them with the stimuli he used, such as food and the ringing of a bell, from behind a screen, and then see how much saliva they had produced as a result.

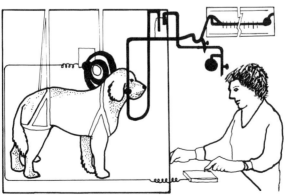

stimuli as a result of reward. Animals tend to repeat rewarded actions and reward can thus be used to alter their behaviour. The dove in the Skinner box referred to in the last chapter soon learns that pecking the right-hand key gives it water while the left-hand key yields food. Based on ideas such as these, elaborate theories of animal learning have been built up which stress the role of the environment and of experience, rather than genetics and inheritance, in the development of behaviour. At an extreme, such psychologists have suggested that learning in all animals is subject to the same rules and that animals do not have predispositions about the sort of things they will learn but will build up any associations with equal ease provided that their sensory and motor systems allow them to do so. Obviously one would not attempt to train a sparrow to fly in the dark or a tortoise to walk on its hind legs.

The following famous assertion of Watson's illustrates the fervour of his belief that the environment rather than heredity was the major determinant of behaviour: 'Give me a dozen healthy infants, well-formed, and my own specified world to bring them up in and I'll guarantee to take any one at random and train him to become any type of specialist I might suggest – doctor, lawyer, artist, merchant-chief and, yes, even beggar-man and thief, regardless of his talents, penchants, tendencies, abilities, vocations, race of his ancestors.' He went on to admit that he was overstating his case, but insisted he was doing so as a counter to the sweeping claims of his opponents.

Both views such as this and those of the ethologists could not be right,

Figure 5.2. Conditioning involves the formation of new connections. Before it takes place, a stimulus leads to a response. As a result of that stimulus being associated with a different one (the conditioning stimulus), the second stimulus comes to lead to the response. In the case studied by Pavlov, the bell came to lead to salivation without the need for food to be present.

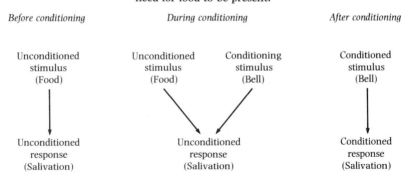

but to some extent they existed because their adherents were studying different things. Psychologists were interested in learning and in behaviour patterns, like the lever pressing of a rat in a Skinner box, which could be modified by it so leading an animal to behave differently from other members of its species. By contrast, the ethologists were interested in more fixed behaviour, such as courtship displays, which were common to all members of a species. Why did they consider such behaviour to be 'innate' and how does the evidence they used hold up to detailed examination?

5.2 Ethologists and instinct

In 1828 a boy of about 16 was found in the market-place at Nuremberg in Germany. He could not speak and his behaviour was like that of a little child, so he was labelled 'the wild boy'. His name was Kaspar Hauser and, when he was able to communicate, he explained that he had been brought up entirely in isolation by a man who had kept him and cared for him all by himself in a hole. Kaspar Hauser may well have been a fraud, but he has given his name to the deprivation experiments which have often been used by ethologists in an attempt to understand whether or not the behaviour they were studying was innate. The idea was to deprive an animal of relevant aspects of experience and, if the behaviour appeared despite this, it could be regarded as inherited whereas, if it did not appear or developed abnormally, it could be assumed that the missing experience was important.

There are many striking examples of behaviour appearing apparently normally in animals whose experience is distinctly abnormal. A case in point was shown by experiments carried out by Janet Kear on ducklings which she had reared by hand from the egg. Most duck species nest on the ground, but some of them, such as the wood duck, do so in holes far up in trees from which the fledglings have to leap to the ground beneath (Figure 5.3). Fortunately they are light and fluffy when they do so and, as a result, they bounce and run off rather than suffering multiple fractures. Kear tested chicks of various species on a visual cliff such as that shown in Figure 5.4. In this apparatus the animal is placed in the centre and can move either to one side where there is a drop looking like a cliff beneath the glass or to the other where the floor is immediately under the glass so that it looks shallow. The behaviour of the ducklings was appropriate to their normal nesting place. Tree-nesters did not avoid the deep side of the cliff but, if they moved that way, they would leap as if casting themselves into space. On the other hand, the ground-nesters

Figure 5.3. Holes high in trees provide safe nesting places for some ducks, but their fledglings must leap from there to the ground. Unlike other ducklings they are not afraid of heights. This goldeneye nest cavity was made by a black woodpecker. (Photograph by H. Dow, Nature Conservancy Council.)

tended to move to the shallow side rather than the deep one, suggesting that they avoided heights. Interestingly, if they did move to the deep side their behaviour was quite different: they pushed off with both feet as they would when moving out from the edge of a pond!

Being hand-reared, these young birds had had no opportunity to learn the actions they showed from others or from earlier experience with ponds or with cliffs. Many ethologists would therefore have used this evidence to argue that the behaviour must be innate because it develops despite deprivation of opportunities for learning, and the main motivation for carrying out such experiments has often been to discover whether behaviour is 'innate' or 'learnt'. Sometimes, as with the ducklings, behaviour develops normally even though the animal is reared in a very impoverished environment. Another good example here is the hoarding behaviour of squirrels whereby, even in captivity, they will bury nuts underground to form stores which they eat later when food is scarce. If such an animal is raised on a liquid diet, so that it never experiences nuts, with masses of this food available the whole time, so that it never needs to hoard, and on a bare floor so that digging is impossible, the first time it encounters nuts and earth it still digs a hole and buries them.

By contrast with these experiments, others have shown behaviour patterns to be radically altered unless particular experiences are available. If a young dog is reared in total isolation from all others, in a chamber in which it can be fed and cared for without contact with even its human caretakers (Figure 5.5), it turns into a strange animal, apparently careless

Figure 5.4. A visual cliff, as used by Kear in her studies of ducklings. The whole apparatus is covered by a sheet of glass, but the left side appears to be shallow while that on the right looks deep.

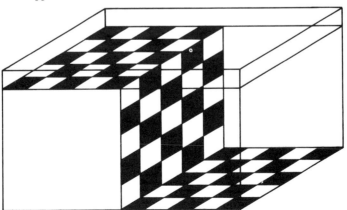

of its own welfare. It will put its paw in a fire and singe it, and it will repeatedly approach an object that gives it an electric shock. This is quite unlike a normal puppy of the same age, which hastily withdraws from such painful experiences. Furthermore, the normal puppy will yelp when hurt, whereas the one that had been isolated behaves as if it did not even feel the pain. Clearly its deprivation has led to a drastic alteration in the way its behaviour developed. In another example, to which we will return later, a young male chaffinch reared out of earshot of all birds of its species has been found to develop a very simple and unstructured song quite unlike that of a normal adult. If he is deafened as well, so that he cannot even hear his own efforts at singing, the song he produces is even worse, being little more than a screech.

These Kaspar Hauser experiments vary enormously in what they actually deny the animals. The isolated chaffinch cannot copy song from other birds, but he can practice singing; if deafened he can still practice, but cannot hear the outcome. The hoarding squirrel has had experience

Figure 5.5. Studies of behavioural development in dogs involved rearing puppies in isolation both from other dogs and from humans. They were kept in chambers with two doors and a partition so that one part could be cleaned out while they were in the other. As a result, they were isolated from all contact with the outside. (After W. R. Thompson & R. Melzack (1956), *Scient. Amer.* January, **194**.)

of neither nuts nor earth so its deprivation is more extreme: it cannot copy from others, it cannot learn for itself, not can it practice digging in earth, though it can carry out the movements concerned on the bare cage floor. In many cases it has proved very difficult to deny animals all the experiences one might think likely to be relevant. Finches with no hay that they can use for nest building will carry seed, lettuce, faeces and feathers to their nest site. Some bizarre behaviour results when the feathers used are still attached to themselves or to their mates. A bird may have to walk rather than fly to its nest site because it is holding its own wing in its beak as nest material; after carefully placing the wing in a corner of the nest it will then fly down, pick the wing up again and struggle back to the nest with it!

5.3 A false dichotomy

To ethologists such as Konrad Lorenz, deprivation experiments were the main evidence used in deciding whether or not particular aspects of behaviour were innate. But the examples mentioned above point to a problem in interpreting these experiments: it is almost impossible to tell exactly of what one has deprived the animal. This difficulty was one of those pointed out forcefully by a leading critic of Lorenz's theories, the American psychologist Danny Lehrman. The points he raised deserve discussion here because they were very influential in leading to current views of behaviour development. They fall under three main headings.

5.3.1 *Criticism of the deprivation experiment*

Lehrman argued that deprivation could show where experience was important but not where it was not. Thus the chaffinch which sings abnormally after being isolated shows that hearing the song of others is essential for normal song development. But the bird that builds a normal nest the first time it encounters hay may have had many experiences, from grooming its own feathers to husking seeds, which could have contributed to its nest building capability. Sometimes the strangest and most unexpected experiences turn out to be important. Thus baby rats will not urinate for the first time unless their genital area is stimulated: normally this occurs because their mothers lick them soon after they are born, but if they are isolated before this occurs they will swell up until their bladders burst. Given the existence of such unlikely influences, it is rash to assume that a simple deprivation experiment has removed all the experiences that may be relevant.

5.3.2 Criticism of viewing learning as the only environmental influence

The environment has many influences on animal development to which it would not be appropriate to apply the word learning. The example of the baby rats can also be used to illustrate this point. Though the young rat is stimulated to start urinating, no one would suggest that it has learnt to do so.

Figure 5.6. The kittens studied by Blakemore & Cooper (1970, *Nature* **228**, 477–8) obtained all their visual experience in tubes lined with either vertical or horizontal stripes (*a*). As a result the units in their visual cortex were found to be sensitive only to lines in orientations similar to that they had experienced (*b*). Each line in (*b*) represents the preferred orientation of a cell in the visual cortex of one of the kittens: an animal reared with horizontal stripes on the left, with vertical on the right.

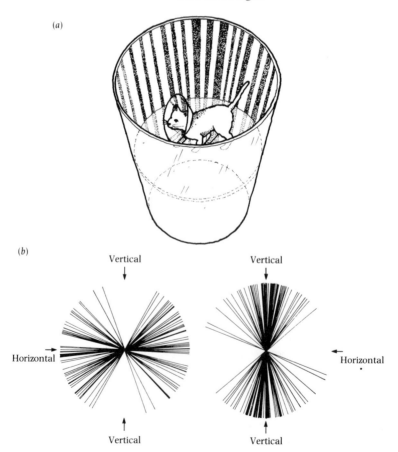

Lehrman himself cited another interesting example of development being influenced by the environment in an unexpected way. This was the effect of dilute magnesium chloride solution on eye development in the common killifish. Normally these fish develop, as indeed do most vertebrates, with one eye on either side of their head. However, if reared in magnesium chloride, a small proportion form only a single, centrally located, eye. This certainly is not learning, but it is a very substantial change in the animal's structure, which doubtless affects its behaviour in many ways, and is brought about by a single simple environmental influence.

Many other environmental influences which affect behaviour have been described. Kittens growing up in an enclosure covered with vertical stripes become incapable of seeing horizontal ones, and those reared with the opposite orientation show the reverse effect (Figure 5.6). In toads, the visual areas on the two sides of the brain are wired together so that cells seeing the same point in space become connected to each other as a result of their common experience rather than by some automatic programme. As toads can reach very different sizes, this strategy of developing connections is a good one if they are to achieve the perfect binocular vision which is essential to them for prey capture.

These examples all concern environmental inputs which will, directly or indirectly, have an effect on animal behaviour but none of which would be called learning. In some cases, like the failure of young rats to urinate unless they are licked, the effect is quite specific. In other instances, as where an abnormal environment leads to a visual deficit, a much more general influence on numerous different aspects of behaviour is bound to occur. Many such environmental effects are now known, and those such as nutritional deficits and oxygen shortages at particular stages of nervous system development are examples of ones known to have a profound effect on behaviour.

5.3.3 Criticism of viewing behaviour as either/or

The discovery that many unexpected factors can influence behaviour has led ethologists to examine the development of behaviour much more carefully and to avoid making sweeping statements about how this or that behaviour pattern is 'innate' or 'inherited'. These words suggest that the behaviour is absolutely fixed and that it would develop in exactly the same way no matter what environment the animal found itself in. Konrad Lorenz clearly felt that many behaviour patterns were like this as he referred to them as being 'blueprinted in the genes'. But, as Lehrman pointed out, expressions like this give a false impression of behaviour development. By depriving the animal of certain influences, one can show that they are unnecessary for its behaviour to develop normally. But where should one stop? There are obviously some very specific effects one might predict as being important to a particular behaviour. For instance, experience with nest material or seeing another animal nest building might be two that could well be essential to normal nest building development. However, there is a continuum of possible influences from ones such as these to very unlikely ones, like experiences during feeding or grooming, which do after all, like nest building, involve manipulating things in the mouth. One cannot be sure that experiences such as these have no importance for nest building. Ultimately some environmental influence is bound to be important in the case of every behaviour pattern because behaviour does not exist in the genes and the genes have to have an environment in a particular temperature range, supplied with oxygen and with nutrients, and so on before the information they contain can be read out to make an animal.

An important point to appreciate here is that natural selection has only determined how development should take place in the normal environment of each species. Thus a behaviour pattern may appear extremely fixed and constant in all individuals of a species because their genes interact with that environment to ensure that this is the case. However, moved to another environment, different from any the species has encountered before, the result may be quite different. Natural selection cannot ensure that the genes will interact with any environment to give the same result; it can only give the right outcome in environments in which it has had a chance to work. Hence fish in magnesium chloride solution may well develop only a single eye. Likewise domestic animals, or for that matter humans, may develop quite differently from the way they would have done in nature. This can lead to behavioural problems, but it is also a point of optimism: even if natural selection favoured individuals that were highly aggressive, this is no reason to think that we are irrevocably committed to being so. Given a different environment from that in which evolution took place, the same genes may produce individuals with a quite different mix of characteristics which could easily be much less aggressive.

On the other side of the dichotomy, the findings of ethologists have also had a marked impact on how psychologists view their subject. For, of course, it is equally sweeping to refer to a behaviour pattern as totally environmentally determined as if the animal's genes were of no significance whatsoever. As psychologists have studied learning in a wider diversity of animals they have found that different species have different capabilities and that they find some tasks easy to learn while others, which might seem similar at first sight, prove very difficult. Such results are often referred to as demonstrating 'constraints on learning'. They indicate that heredity has a profound impact on what tasks the animal is capable of learning to perform.

Most psychologists interested in learning still study rats and pigeons but, even within these species, some interesting constraints have been found. It is easy to train a rat to press a bar to obtain food but not to avoid an electric shock, mainly because the natural reaction of a frightened rat is to 'freeze' and this is incompatible with bar pressing. In some instances where training was successful the rat was found to be freezing on top of the bar rather than pressing it with a paw! In another experiment, shown in Figure 5.7, it was found easy to get rats to associate a taste with sickness and a sound with pain but hard to train them to make the opposite pairings. To an ethologist this result might not seem surprising: in nature, tastes

Figure 5.7. Rats that have drunk sweet water and become sick avoid that water again, and rats in which a clicking sound is followed by electric shock show defensive behaviour after they hear the click. But rats do not easily learn to associate a clicking sound with nausea when they drink, or sweet water with electric shock. (After J. Garcia, J. C. Clarke & W. G. Hankins (1973), in *Perspectives in Ethology*, ed. P. P. G. Bateson & P. H. Klopfer, pp. 1–41, Plenum Press, New York.)

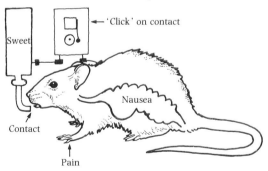

		Consequences	
		Nausea	Pain
Stimulus	Sweet water	Avoids	Does not
	'Clicking' water	Does not	Avoids

are more likely to be a good indication of which food is bad for you than are sounds, whereas pain is more likely to be experienced alongside sounds such as those made by the approach of a predator. But results such as these were something of a surprise to those psychologists who were committed to the idea that an animal's heredity was of no significance and that it should therefore make various different associations with equal ease.

The recent results of both ethologists and psychologists have thus led to a coming together of the two fields as far as development is concerned. Most would now agree with Donald Hebb's remark that both genes and environment are 100% important in the development of all behaviour: no genes, no behaviour; no environment, no behaviour. And, as we shall see in the rest of this chapter, some of the interactions between the two are fascinating and subtle. It certainly does not do justice to them to ask simple questions about whether behaviour is genetically or environmentally determined.

This argument does not mean that carrying out deprivation experiments is a useless enterprise. Depriving an animal of particular experiences can indicate whether or not they are crucial for normal development. Furthermore, deprivation experiments have shown that some aspects of behaviour are much more flexible than others. Certain actions, like hoarding in squirrels, develop in much the same manner even though the environment is changed substantially in various different ways. Others are altered by even minor manipulations. Thus some behaviour patterns are stable and some labile, and it is interesting to examine the reasons for these different developmental strategies. The consideration of a few case histories will enable us to do this, at the same time as illustrating how ethologists have studied the development of behaviour now that the word 'innate' has all but disappeared from their vocabulary.

5.4 Some case histories

5.4.1 *Song development in birds*

Bird song is the classic example of how both genes and environment have a crucial role to play in development. Since the pioneering work of W. H. Thorpe on chaffinches, many species have been studied and it has become clear that learning plays an important role in all of them and also that there are constraints on what they are able to learn.

Thorpe was able to show, by a series of experiments on hand-reared chicks, that chaffinches learn from other chaffinches. As in most other species, only the males sing. Thorpe found that, if he raised young males

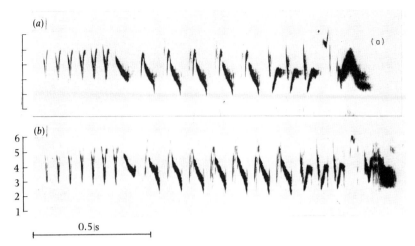

Figure 5.8. Birds often learn the sounds that they produce from other individuals very accurately. Just how well they do so can be assessed by examining the traces made by a sound spectrograph, as shown here. This machine produces a plot of frequency against time, the darkness of the plot showing how much energy was present at that particular moment and pitch. (*a*) The song phrase of a chaffinch played to a young individual when it was starting to sing and (*b*) the copy that it produced. (After P. J. B. Slater & S. A. Ince (1982), *Ibis* **124**, 21–26.)

in total isolation from all others, the song they produced was quite different from that of a normal adult. It was about the right length and in the correct frequency range; it was also split up into a series of notes as it should be. But these notes lacked the detailed structure found in wild birds, and the song was not split up into distinct phrases as is usually the case. This suggested that song development requires some social influence. Later experiments in which young birds were played recordings of songs showed just how precise this influence was: many of them would learn the exact pattern of the recording they had heard (Figure 5.8*a*, *b*). A remarkable feature here was that birds were able to copy precisely songs that they only heard in the first few weeks of life, yet they did not sing themselves until about 8 months old. They are thus able to store a memory of the sound within their brain and then match their own output to their recollection of it when they mature.

Young chaffinches normally learn only chaffinch song, though Thorpe found they could be trained to sing the song of a tree pipit, which is very similar to that of their own species. In general, however, the constraints on bird's learning ensure that they only learn songs appropriate to the species to which they themselves belong. These constraints may be in their

Fig. 5.8 (*continued*)

(*c*) The phrase 'Learning what to say' as spoken by the human voice
and (*d*) as imitated by an Indian hill mynah. (After W. H. Thorpe
(1967), *Proc. int. Orn. Congr.* **14**, 245–63; original sonagrams by
courtesy of Mrs J. Hall-Craggs.)

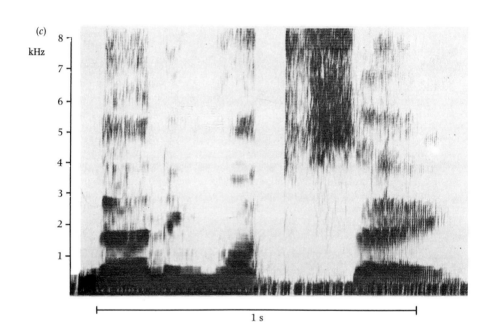

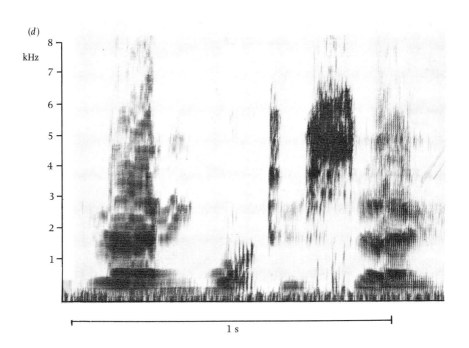

(*d*)

neural circuitry, the young bird hatching with a rough idea of the sounds that it should copy. The crude song of a bird reared in isolation gives some clues as to what this rough idea may be: the length, the frequency range and the breaking up into notes are all aspects of chaffinch song shared between normal birds and those reared in isolation. In other cases the constraints seem to be more social, young birds being prepared to learn from the individuals that reared them or with whom they have social interactions. Whatever the nature of these learning rules, there is no doubt that they are effective: it is very unusual to hear a wild bird singing a song which is not typical of its own species despite the many different songs which often occur in a small patch of woodland.

Not all birds show a learning pattern which is the same as that of chaffinches. There are some species which produce normal sounds even if deafened, so that they cannot hear their own efforts far less copy those of others. The cooing of doves and the crowing of cocks are examples here. In other cases, such as parrots and hill mynahs, birds can be trained to copy a huge variety of sounds (Figure 5.8c, d), though those they learn in the wild are usually more restricted. The amazing capability of mynahs has apparently arisen simply because birds in an area learn a small number of their calls from each other, males from males and females from females. The ability to master these notes has led them, incidentally, to be capable of saying 'hello', copying wolf-whistles and mimicking the sound of a car door shutting. The capacity for mimicry in some birds can be just as remarkable in the wild and shows that not all of them are as constrained as the chaffinch in what they will copy. The starling and the mocking bird are perhaps the best known examples but a less well-known species, the marsh warbler, must take the prize. Males of this small migrant learn sounds from many other birds as they move down from Europe into Egypt and then on to East Africa in the first autumn of their lives. They then string them together into an amazingly varied song on their return north again the next spring. The average number of species that a male imitates is 76: the only sounds he seems to reject are the really deep ones which he cannot master because of his small size!

The timing of learning also varies a great deal between species. In chaffinches it is restricted entirely to the first year of life, and adult males that have not learnt a normal song by then cannot be trained to produce one thereafter. The male sex hormone testosterone, which induces them to start singing, seems also to stop them learning more. Their learning thus occurs in a short 'sensitive period' early in life. Such sensitive periods are a common phenomenon in behaviour development, but they are not always found. While some birds may only learn as fledglings or as young adults or, as in the chaffinch, in both these periods, other species carry on learning throughout life. An example here is the canary: males in this species modify their song from year to year regardless of their age.

Bird song, then, provides a fine range of examples of behaviour development, some of which are tightly constrained and others highly labile. But, despite the variations they show, in no case is it reasonable to think of development as entirely either genetically or environmentally determined.

5.4.2 *Recognising predators and prey*

All animals must distinguish between things that they can eat and things that they cannot; they must also recognise other animals that are likely to eat them. These tasks may be quite simple if the animal specialises on one or a few types of food, and if there are not many different predators that could eat it. On the other hand, species like rats and humans eat a wide variety of different foods and must distinguish those that are nutritious from those that are poisonous. Threats too can take many different forms: a killer whale, a polar bear and a man with a gun have little in common to look at but are all potentially lethal to a seal.

The extent to which food preferences are present at birth varies considerably between species. Newborn snakes show strong species differences in their responses to various food items which are in line with the preferences of adults. Even within a species, populations may differ, as Figure 5.9 shows.

In other cases, preferences can be acquired, and this enables animals to adjust their feeding to match availability, a skill at which rats are past masters. Rats can learn about food for themselves or they may pick up information from others. If they encounter a new type of food, they will have a brief nibble and then wait for a period before returning to it. Should they feel ill during this time they will not come back but will reject the food thereafter. This is an excellent way of avoiding being poisoned, though it is infuriating to those humans who are bent on exterminating rats.

Rats seem to learn about food from others in a variety of ways. One route, curiously enough, is through their mother's milk. Young rats, when newly weaned, show a preference for the foods on which their mother had been feeding during lactation. Doubtless chemical cues about the nature of the food pass to them through the milk and help them to learn by her experience. Later in life, when living in their social groups, rats are also known to prefer foods on which one of their companions has been feeding.

Odours clinging to the fur or detected in the faeces of the one that has found the food seem to be involved.

Eating food that is not very nutritious or makes one a bit sick is not a disaster, whereas a brush with a predator may be one's last. Again, evolution has led to various different developmental schemes which avoid this eventuality. In birds, some species appear to respond to specific features of predators such as hawks and owls, without prior experience of them, while others take evasive action from all large objects that loom up to them and only slowly start to ignore those that prove to be harmless. Some neat experiments in this area have been those by Vilmos Csányi, a Hungarian ethologist, on paradise fish. Young fish of this species will swim towards and investigate either a pike or a goldfish introduced into their tank (Figure

Figure 5.9. In California, the garter snake *Thamnophis elegans* feeds almost exclusively on slugs near the sea, but mainly on frogs and fish inland where slugs do not occur. Young snakes offered one piece of slug each day for 10 days show similar preferences, those from two coastal areas tending to accept them, while inland ones almost invariably reject them. These preferences do not result from experience as none of the snakes had encountered slugs before the tests began. (From S. J. Arnold (1981), *Evolution* **35**, 510–15.)

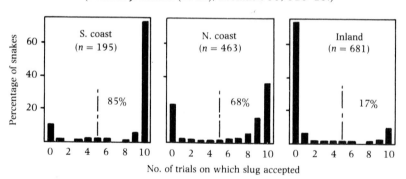

5.10): a thoroughly dangerous enterprise in the case of the pike as it is likely to chase and catch them. If they are chased by it, they will avoid approaching on later occasions while still being prepared to swim close to the goldfish. However, a mild electric shock the first time they meet a goldfish makes them treat it with caution too. Thus these experiments illustrate a generalised response to various different objects at the outset which does not place the animal at too great a risk but predisposes it to learn which are dangerous and which are not. No matter what mix of predators there are in its environment, its experience enables the young animal to recognise them and take evasive action while continuing to feed peacefully when harmless animals pass by.

5.4.3 Social development

Most animals do not have an entirely solitary existence, but must develop social relationships with other individuals of their species. Initially this may just be the mother or father to whom an attachment grows early in life ensuring that the young animal stays close by and does not wander into danger. In birds, this process is called imprinting. The young chick does not immediately recognise its mother and imprint upon her, but will follow and thus imprint upon any large moving object that it is exposed to during a sensitive period in the first few days of life, be this a model chicken, a toy car or, in the most famous example, Konrad Lorenz (Figure 5.11). The chick may be more fussy about sounds, and there is some evidence that chickens prefer objects that cluck and ducklings those that quack, but generally its visual preference is, at least initially, for the largest and most striking thing around. Flashing lights are especially good. By following whatever object it has encountered for a few days, it learns all about it and then seeks it when distressed and rejects other objects even when these look similar. If imprinted on a red watering can, it comes to prefer this to a blue one or to a red jug. In nature, of course, the result of this process

Figure 5.10. Young paradise fish do not avoid pike at first, but learn to do so as a result of being chased by them (From V. Csányi (1985), *Behaviour* **92**, 227–40).

is that it learns what its mother looks like; red watering cans do not normally wander round the territories of most birds.

In mammals too there is a process analogous to imprinting which leads to young ones being attached to their mothers. Sigmund Freud thought that this occurred through 'cupboard love', the mother rewarding the young one with milk so that it benefited from staying close to her. This is not so, however. In a classic series of experiments, Harry Harlow studied young rhesus monkeys reared with two mother substitutes (Figure 5.12). One of these was a frame covered with towelling to make it cuddly, the other was the frame alone but with a bottle of milk mounted in it. When

Figure 5.11. Young chicks follow the first moving object that they see after hatching and become attached to it. In the wild this object will invariably be the mother, but laboratory experiments show that chicks will become attached to objects that look very different if they are exposed to them at this time. The chick shown here is walking in a wheel towards a rotating box with a red flashing light inside it: a very effective stimulus for imprinting (photograph by B. McCabe.)

frightened the young animals rushed to the towelling mother and ignored that which provided milk. Indeed, if the two objects were close together, the infants learnt to cling to the cuddly one and lean across to drink from the other.

Rhesus monkeys therefore become attached to their mothers without the need to be fed by them, and they use them as bases from which to explore. The parallels between this growth of attachment and the imprinting of young birds are very interesting. Human babies, when they first start to smile at a few weeks old, will do so to any face that appears and even to pictures of faces with the features all jumbled up (Figure 5.13). Later the features must be arranged correctly, and later still the face must be that of someone whom they know well, and often only that of their mother will do. Any stranger who has lifted a happy smiling 9 month old from its mother's arms is likely to know the dramatic effect this can have! Thus babies, like birds, start off being responsive to a wide range of stimuli, but later become attached to those of which they have much experience and frightened by all others.

Baby monkeys spend a lot of time clinging to their mothers and suckling, but as they become more mobile they begin to move about, exploring their world, meeting different adults and playing with their age-mates. The mother is close by at first, keeping an anxious eye and grasping the infant if danger looms. But later the infant wanders further afield, only occasionally returning to her to feed and when in need of comfort. Interestingly, now their roles reverse: instead of the mother grasping the infant and restraining its wanderings, she tends to push it away and reject its approaches, as if encouraging its independence. This is a well-known

Figure 5.12. The two models used by Harlow to study the attachment behaviour of young Rhesus monkeys.

phenomenon called 'weaning conflict'. The infant wants to go on feeding at the breast; the mother wishes to stop it doing so (and who could blame her as the young one's teeth erupt one by one!). That the interests of mother and infant should diverge in this way seemed curious until a few years ago, but this is one aspect of behaviour which has come to be better understood thanks to recent advances in our understanding of how evolution works, a topic to which we shall return in Chapter 7. These ideas suggest that there should indeed be a period when the mother would be best to wean her young one, while it is to the advantage of the infant to persist in being fed.

5.5 Conclusion

The development of behaviour is clearly a complex process and many different strategies have been described. Some of these are more flexible than others, so certain behaviour patterns develop very similarly in a wide range of environments, while others are much more affected by environmental differences. Ethologists have always been especially interested in animal signals, such as courtship and aggressive displays. If these are to be understood by others, animals must get them right and thus they tend to be very similar throughout a species: hence the belief that they were fixed and 'innate'. Psychologists, with their very different interests, saw little fixity in what they studied. But now, with so many constraints

Figure 5.13. At first young babies are interested by a wide variety of stimuli like the two shown here, but later they only attend to that which looks like a face. Later still, when they have experienced many real human faces, they do not take interest in either of these simple pictures. (After J. Kagan (1970), *Science* **170**, 826–32.)

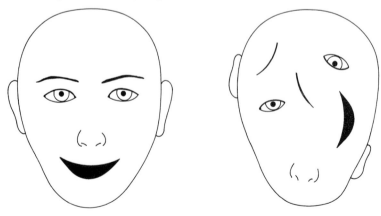

on learning demonstrated on the one side, and with a great many environmental influences, including learning, shown to affect even the most constant of behaviour patterns, there is no longer this stark contrast of viewpoint. Genes and environment affect all behaviour and, as we have seen from many of the case histories in this chapter, discovering the way they interact with each other to produce an adaptive outcome has been one of the most interesting fields of animal behaviour in recent years.

6

Evolution

Behaviour leaves no fossils, though sometimes it is possible to surmise how animals in the past must have gone about their business. An animal with wings is likely to have been able to fly, and fossilised soldier termites will undoubtedly have defended their nests, as do their equivalents today. But such clues are rare and generally we must rely on studying how animals today behave to give us an idea about the evolutionary history of their behaviour. This is a hazardous business, and firm conclusions are hard to come by, but some of the approaches that have been used provide evidence about how evolution must have moulded behaviour. Before considering these, however, it is as well to say something of the genetic basis of behaviour, for without a genetic basis evolution would simply be impossible.

6.1 Behaviour genetics

In the last chapter it was argued that all behaviour is dependent on heredity, but the fact that genes affect a behaviour pattern is not sufficient for evolution to take place. Evolution acts because of differences in the success of individuals: those doing better leave more of their genes to the next generation than those doing worse, so that fit genes spread and the less fit become scarcer. But as it is the genes that pass from one generation to the next, such changes in frequency can only occur if the differences between individuals are based on differences in their genes. Otherwise there is no way that the better behaviour can be passed on to its offspring by its possessor. Strictly speaking then, evolution requires not just that behaviour has some genetic basis, which all behaviour does, but that variations in behaviour are based in part on genetic variations on which selection can act.

Studies of the effects of genes on behaviour have taken several forms. Two of these will be discussed here: the examination of mutants and the use of selection experiments.

6.1.1 Studies of mutants

By breeding from mutant individuals it is possible to generate a stock which is the same as that from which it is derived except for a single gene locus. An example is the *Yellow* mutant in the fruit-fly *Drosophila melanogaster*, so called because of the colour change it induces in its possessors. Margaret Bastock, in a famous study, showed that males with this gene were slower

Figure 6.1. Some of the movements involved in the courtship of *Drosophila melanogaster*: (*a*) orientation of the male towards the female; (*b*) scissoring, in which he moves both his wings in and out to the side; (*c*) vibration, which involves the wing nearest the female being moved out and vibrated to and fro; (*d*) licking of the female's genital area by the male and, finally, (*e*) copulation. (From A. Manning (1965), *Viewpoints in Biology* **4**, 125–69.)

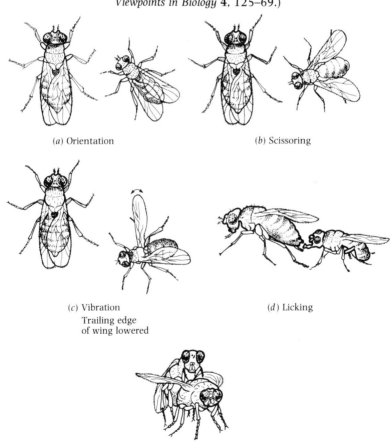

(*a*) Orientation   (*b*) Scissoring

(*c*) Vibration   (*d*) Licking
Trailing edge
of wing lowered

(*e*) Copulating pair

at achieving mating than normal males and, on examining their courtship she found that they displayed less of a component called vibration (Figure 6.1), which serves to stimulate the female. This single gene does therefore influence behaviour, in this case by altering frequency. Such an effect may not appear very substantial but, of course, if the frequency of a behaviour pattern is modified down to zero it will not appear at all.

Later studies have shown just how complex are the relationships between genes and behaviour. A good example here is the way in which single gene mutations affect the activity of mice in a running wheel (Figure 6.2). Examination of 31 different mutations, chosen because their presence altered coat colour and so was easily spotted, showed that 12 of them also altered the activity of their bearers. As there is no reason to think that coat colour and activity are like to be linked to one another, this result suggests that about 40% of all genes probably influence activity in one way or another. This is certainly a far cry from the idea that one gene produces one behaviour pattern but, if one thinks about it, it is only to be expected. A feature such as activity is likely to be affected by innumerable factors: leg length, visual acuity, lung volume and muscular strength are a few that come easily to mind. These in turn depend on many physiological and biochemical processes and hence on many genes.

Just as one behaviour pattern is influenced by many genes, so the opposite is also the case: one gene may influence many patterns of

Figure 6.2. The activity of mice in running wheels: a behaviour pattern which has been found to be influenced by many different genes.

behaviour. Again this should not be too surprising. Imagine a human gene which alters the level in the bloodstream of the male sex hormone testosterone so that it is much lower than in males lacking this gene. The mutant males are likely to be physically different, with less beard growth and more poorly developed muscles. The latter features alone will probably affect many aspects of their behaviour, but some will be affected more directly, for testosterone itself is known to stimulate areas of the brain responsible for sexual and aggressive behaviour as well as the vocal apparatus to give deepening of the voice. The mutation would thus be very substantial in its effects on behaviour.

All in all, then, the study of mutants has shown just how complex is the interplay between genes and behaviour. Many genes affect each behaviour pattern and each gene influences many actions. These points alone make it likely that there will be plenty of genetic variability on which natural selection can act. Studies using artificial selection confirm this point.

6.1.2 Selection experiments

If a behaviour pattern can be altered by artificial selection, this indicates that variations in it have a genetic basis on which selection can act and that natural populations are variable with respect to these genes. Many such experiments have been carried out and they have shown, first, that selection can change a wide variety of behaviour patterns and, secondly, just how these changes are brought about.

As in many other aspects of genetics, the fruit-fly *Drosophila melanogaster* has been specially useful here because a large number of generations of selective breeding can be achieved in a short period of time. Aubrey Manning selected for high and low mating speed in fruit-flies. Out of 50 pairs he set up a line from the 10 that mated fastest and another from the 10 that mated slowest. In the fast line he always bred only from the 20% that mated most quickly in each generation and in the slow line he selected the 20% that mated slowest, carrying on for many generations. The maximum effect was reached after just 7 generations, with the fast line averaging 3 minutes to mate while the slow line took 80 minutes (Figure 6.3). Thus selection had had a dramatic effect on mating speed: but how had this been achieved? Examination of the behaviour of the flies showed that several changes were involved. The main effect in the fast flies had been to make them much less active, the males settling down quickly to vigorous courtship while the females showed greater responsiveness. In the slow line, by contrast, both sexes showed prolonged activity after being

placed together and they often rushed past each other without attempting
to start any interaction.

Artificial selection can sometimes give rather surprising results. Arthur
Ewing used a tube with a series of small holes in it in an attempt to select
for high activity in fruit-flies (Figure 6.4). The holes were arranged in
funnel shapes, rather like the entrances to lobster pots, so that a fly passing
through one was extremely unlikely to return again. Ewing then set up
one line by breeding from the flies that passed through the whole system
most quickly and another from those that set off most slowly. The main
effect he found was a decrease in the speed of the slow line but, curiously,
these flies were just as active as the others if placed in an open arena. What
he had succeeded in obtaining was not lower activity, but 'claustrophobic'
flies that disliked passing through small holes. Selection experiments can
often have rather unexpected results such as this, for, as pointed out above,

Figure 6.3. The results of the first nine generations of selection for
fast and slow mating speed in *Drosophila melanogaster*. The points
shown are the mean mating times for two fast lines (×, ●) and two
slow ones (△, ○), showing first the parental times (P1). Note that the
time scale is logarithmic, so that the fast line ended up mating on
average about ten times quicker than the slow. This difference
persisted without any major further change in subsequent generations.
(After A. Manning (1961), *Anim. Behav.* **9**, 82–92.)

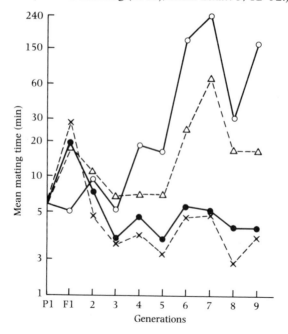

a single aspect of behaviour is influenced by many genes and artificial selection may be effective by altering the frequency of any of them.

The higher the proportion of behavioural variability that is based on genetic differences, the more likely is selection to be effective in altering a characteristic. But it is hard to conceive of a behaviour pattern in which there is no genetic variation on which selection could act. A good example here, which produced results which were surprising at the time, was experiments involving selection for learning ability in rats. Rats can be taught to run mazes for a reward of food at the far end, and some of them learn more quickly than others. Breeding from those that learn best and those that learn worst leads to a progressive divergence over the course of several generations (Figure 6.5). This shows that differences in learning ability amongst rats are to some extent genetically based, otherwise selection could not accentuate them from generation to generation.

These selection experiments show that those studying behaviour genetics are able to select for virtually any characteristic they care to choose, although this is not always changed in a way that one might expect. The ease with which they can do so depends on the extent to which variation in behaviour has a genetic basis; the way selection acts is to alter those features for which genetic variation is greatest. But from the point of view of this present chapter the important message is simple: there is ample evidence of variations in behaviour being affected by genetic variations on which natural selection could act.

6.2 Comparative evidence

Although we cannot go back in time to see how behaviour evolved, we can look at different species alive today and ask how their behaviour has diverged since they split off from one another. This is obviously most

Figure 6.4. Apparatus used by Ewing ((1963), *Anim. Behav.* 11, 369–78) to select for spontaneous activity in *Drosophila melanogaster*. The flies were introduced into compartment *A* and the lines selected were bred in each generation from the last 20% of each sex to leave *A* and the first 20% to reach *F*.

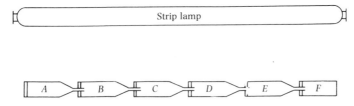

fruitful if the species involved are closely related so that the differences between them are quite slight and the changes involved are reasonably easy to reconstruct. Even then it is not absolutely straightforward for several reasons that hinge on the fact that behaviour is very flexible.

One point here is that behaviour can diverge for reasons totally unconnected with genetics. We have already discussed examples in the last chapter of how behaviour, such as food preferences in rats and song in birds, can be passed from one individual to another by cultural transmission. This can lead to a particular aspect of behaviour being found in one population but not in another, without there necessarily being any genetic differences between the two, a point to which we shall return in our discussion of culture in Chapter 9.

Another problem of interpreting behavioural differences between species is that of convergence. This is always a difficulty in interpreting evidence about evolution, but it is particularly so with behaviour because of its flexibility. Two features may be similar because they are derived from a common ancestor, in which case they are referred to as homologous, or they may be so because their bearers have converged upon the same solution to a common problem, so that they are simply analogous. The wings of birds, the arms of humans and the pectoral fins of fish are homologous, all being examples of the vertebrate forelimb. But the wings

Figure 6.5. Results of an experiment in which rats were bred for brightness and dullness in learning a maze. After six generations, the dull rats were averaging around twice as many mistakes during learning the maze than were those in the bright line. (From W. R. Thompson (1954), *Proc. Ass. Res. nerv. ment. Dis.* **33**, 209–31.)

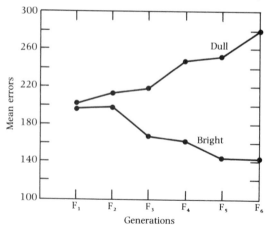

of birds and of insects, while serving similar functions, are only analogous with one another as they are not derived from common structures. In the case of behaviour, distantly related species may have similar characteristics because they rely on similar food or because they live in much the same habitat. Convergence is common and it is not always easy to discriminate between it and similarities due to common ancestry.

The most satisfactory examinations of the evolution of behaviour have involved the study of displays, because they are less prone to the problems that have just been outlined. As one of their prime functions is in species recognition, they tend to be rather fixed and constant throughout a species and yet to differ between species. As we shall see in Chapter 8, their form is often related to the type of habitat to ensure effective communication. However, the differences between them are not so much affected by the environment as happens in many other actions, so convergence is less common and similarities between species are usually proportional to the closeness of their relationship.

Comparisons of displays have helped especially to reconstruct the course of evolution at two different levels: the divergences that have occurred at the level of the family tree, and the more subtle changes that have taken place in the course of two species splitting off from one another.

6.2.1 Family trees

Konrad Lorenz pioneered studying behaviour as a means of making comparisons between species, arguing that it could be used in just the same way as structures to work out the relationships between them. He looked at various ducks and geese and found some behaviour patterns which they

Figure 6.6. 'Burping', a display shown by ducks but not by geese, and thus likely to have evolved since these two groups split off from their common ancestor. It is shown here being given by a pintail. The head is pushed far upwards and the beak is lowered slightly, accompanied by a call which rises in pitch and then falls again as the posture returns to normal. (After K. Lorenz (1941), *J. Ornithologie* **89**, 194–294.)

Figure 6.7. (*a*) Table showing which groups amongst Pelicaniform birds possess various different displays. (*b*) A tree showing the pattern of evolution of the group based upon morphological features with the point at which the displays are likely to have first appeared indicated on it. The distribution of the displays is compatible with the morphological evidence: it is not necessary to make the unlikely assumption that a single display evolved twice. (After G. F. van Tets (1965), *Ornithol. Monogr.* **2**, 1–88.)

(*a*)

	Pelicans	Gannets & Boobies	Cormorants & Darters	Frigate birds
Kink throating (KT)			+	
Head wagging (HW)		+		
Sky pointing (SP) and Hop display (HD) and Wing waving (WW)		+	+	
Bowing (BOW)	+	+	+	
Presenting (PNM) Nest material and Prelanding call (PLC)	+	+	+	+

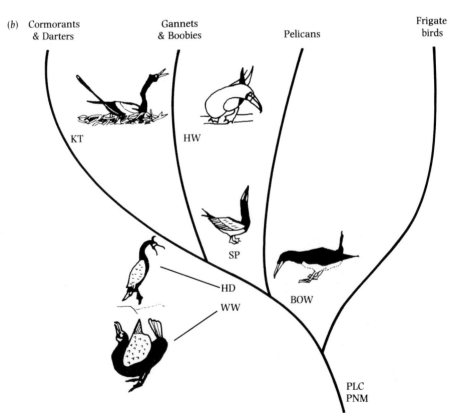

(*b*)

all had in common, like the monosyllabic call of the chick, and others which occurred only in a more restricted group, such as the 'burping' which appears amongst the displays of ducks but not those of geese (Figure 6.6). As displays are often elaborate, it is extremely unlikely that exactly the same one would arise twice during evolution, so one can assume that wing shaking occurred in the common ancestor of all geese and the piping call in that of the whole group. While displays may be lost as well as gained during evolution, the idea behind comparisons such as this is quite simple: two species which share a feature are more likely to be closely related than is either to a third species lacking that characteristic. If many aspects of behaviour are looked at, compelling evidence can accrue about the way in which evolution led to the differences between the species being studied.

A simpler case than that of the ducks and geese is that of the displays of pelicans and their relatives (Figure 6.7), studied by a Dutch ethologist, G. F. van Tets. This is rather a diverse group and their displays differ quite a bit. Some are common to all species, some occur in only one and some in a small subgroup of the species. If one assumes that each display arose only once, then it must have been present in the common ancestor of all the species showing it today. This allows one to draw a family tree suggesting when the species split off from one another. As one would hope if the method was well founded, this tree is the same as that which had previously been suggested on the basis of the structures of these species, confirming the suggestion that behaviour can be a useful guide to evolutionary sequences.

Sometimes animals may have curious behaviour patterns the origin of which it is hard to imagine but, by looking at related species, one can begin to see how they might have arisen. One must be careful, as no living species is the ancestor of any other, but some may have changed less than others since they separated and so may give one an idea of what the ancestral form was like. A striking case here is that of the empidid flies *Hilara sartor*. In this group different species show a wide diversity of related courtship displays. The most perplexing of these is in a species where the male spins a delicate balloon of silk before joining the swarm in which mating takes place (Figure 6.8). In the swarm he presents his balloon to a female and, as if it symbolised a wedding contract, her acceptance of it is a prelude to mating. How on earth could such a bizarre ritual have evolved? E. L. Kessel argued that, if one looks at related species, one begins to see a sequence which could have led to this strange behaviour. There are seven species and they behave in the following ways:

1. The male simply courts the female without giving her any token.

2. The male captures a fly and presents it to the female.
3. The male captures a fly and wraps it in a little silk before giving it to the female.
4. The male captures a fly and parcels it totally in silk before giving it to the female.
5. The male captures a fly and sucks it dry before wrapping it in silk and presenting it to the female.
6. The male picks up a fragment of dead fly and bases a silk parcel upon it.
7. The male spins a balloon of silk without using any prey as a basis.

This series of species suggests a sequence in which changes may have taken place. The first five species feed on insect prey and so those of them that present a gift may be conferring a real benefit on the female. The last two, however, are nectar feeders. Their gifts are of no intrinsic value but evolution has left females with a preference for males that present gifts so, within a swarm, only males that do so are likely to get a mate. The whole procedure has become a ritual.

Courtship feeding is quite common amongst animals and may often help the female to lay more or better eggs so that it benefits both her and the male that fertilises her eggs. A neat trick here is that of a scorpion-fly species in which males sometimes mimic females and approach other males that are bearing their gifts and looking for mates. By putting on all the appearance of being receptive, these mimics are presented with dead flies just as if they were females. Having received their gift they then fly off in search of a female of their own!

Figure 6.8. Before mating (*b*), the male of empidid fly *Hilara sartor* courts the female (*a*) by presenting her with an empty silken balloon as a token.

(*a*)   (*b*)

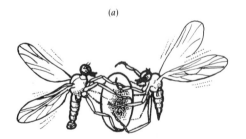

6.2.2 How species diverge

Animal species can be defined as groups of individuals that actually or potentially interbreed with one another. Members of different species tend not to pair with each other in the wild and one of the main reasons for this is that they have distinctive displays which advertise the exact species to which they belong. It is usually highly disadvantageous for animals to mate across species for, even if hybrids are formed, these tend to be either sterile, like mules, or at a disadvantage compared to pure-bred animals. Thus it is in the interests of males to produce clear, species-specific signals and of females to reject those that are not exactly right.

There are several ways in which changes may occur in courtship displays to make species distinct from one another. We can return to fruit-flies to illustrate some of these. When these little flies court, the male orients towards the female and displays to her with wing movements which usually consist of extending the wing nearest to her and vibrating it rapidly (Figure 6.1). This drives a stream of air towards her, which may well bear a stimulating odour, but it also makes a sound which can be heard if a pair of flies is induced to court on top of a sensitive microphone. During evolution, the exact form of this courtship has come to show differences between species in several different ways:

1. Changes in threshold. Two closely related species, *Drosophila simulans* and *D. melanogaster* show this well. When pairs are put together the male *simulans* takes longer to begin courting and appears more sluggish in his actions, but the female is more receptive, so counteracting the effect. Her threshold for accepting the male is thus lower, and she accepts him more rapidly.

2. Change in the senses used. The courtship of *simulans* is slowed down in the dark, whereas that of *melanogaster* is not, suggesting that visual cues are important to the former but not the latter.

3. Changes in the frequency of an action. In *melanogaster* vibration occurs frequently, while scissoring of the wings together is very rarely shown. By contrast, *simulans* shows more scissoring and less vibration.

4. Changes in sequence. Flies of the group to which *melanogaster* and *simulans* belong show these wing movements before mounting, while in the group to which *Drosophila rufa* belongs vibration occurs during mounting.

5. Timing changes. These are best illustrated by the sounds produced by different species. Both the interval between pulses and the frequency of the sound within a pulse differ between species. Interestingly, in cases

where one of these features is the same in two species the other is very different and this is probably sufficient to isolate them from each other (Figure 6.9).

Each of these differences is slight in itself but a combination of them or a succession of changes in one of them can lead to marked contrasts between one species and another. As was mentioned earlier, a display which becomes progressively rarer may eventually cease to appear at all and the presence or absence of behaviour patterns may depend on such gradual shifts.

6.3 The origin of displays

Comparison between closely related species can therefore indicate how gradual changes have taken place during evolution. A more difficult question is where these displays came from in the first place. Presumably animals did not suddenly start leaping up and down and waving their limbs around for no apparent reason, yet actions such as these may be crucial

Figure 6.9. (*a*) Oscilloscope trace of the sound produced by the wing vibrations of a courting male *Drosophila persimilis*. (*b*) Table to show how two features of these sounds vary amongst five closely related species from the group to which *persimilis* belongs. All are distinct from each other in one or other of these measures with the possible exception of *pseudoobscura* and *ambigua*. That these two are so similar is of no consequence to selection as the former is European and the latter North American. (From A. W. Ewing & H. C. Bennet-Clark (1968), *Behaviour* **31**, 288–301.)

(*a*)

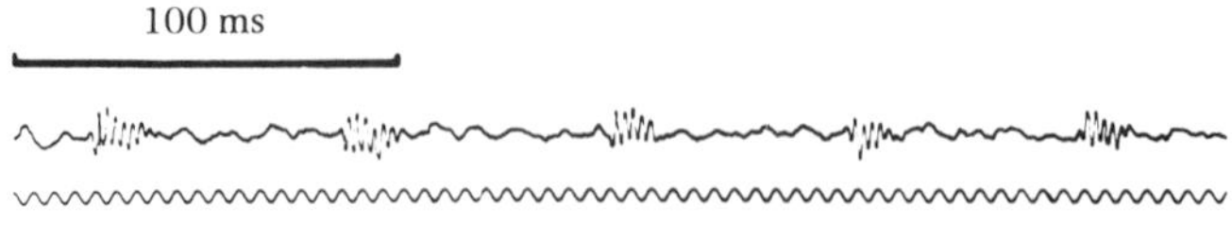

100 ms

(*b*)

Species of Drosophila	Frequency of pulse (cycles per second)	Interval between pulses (milliseconds)
pseudoobscura	258 ± 11	37.9 ± 6.8
persimilis	528 ± 16	53.7 ± 4.3
ambigua	294 ± 35	38.4 ± 3.7
algonquin	226 ± 24	53.3 ± 10.9
athabasca	444 ± 15	11.7 ± 1.1

to their courtship today. Several ideas have been proposed here, based largely on the comparison of displays with the everyday behaviour of the species which show them. If an action looks similar in the two contexts, then perhaps the one was derived from the other and we can ask why this particular act rather than any other should have become incorporated into the display. Of course many displays may be very ancient, so that comparisons may be somewhat tenuous: the wing flapping display of a bird may be derived from the movements made by some ancestral fish with its pectoral fins! Nevertheless, some quite convincing links have been found between displays and other actions.

The following types of action have been suggested as having given rise to displays:

1. Intention movements. Before they perform some action animals often show movements which indicate that they are getting ready to do so. For example, a bird about to take off crouches low on the ground with its muscles braced and its wings out from the body, all set to propel itself up and away. Intention movements are particularly common as a preparation for locomotion, and displays often bear a strong resemblance to them (Figure 6.10*a*, *b*). One reason for this is probably that such actions are common when two animals are close to each other, as they are during

Figure 6.10. The similarity between some display postures and other movements suggests that the former are derived from the latter. Many birds show displays which are probably evolved from intention flight movements. Before taking off, birds crouch low down, with hunched body. They then launch themselves into the air by extending their legs and neck. The two green heron displays shown in (*a*) and (*b*) have obvious similarities with these postures. Some other displays have a clear link with grooming, as in the wing preening display of the mandarin duck (*c*) which is clearly derived from ordinary grooming (see Figure 4.14.)

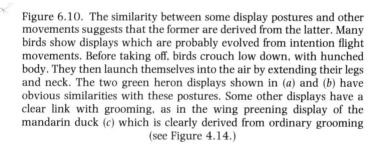

(*a*)   (*b*)   (*c*)

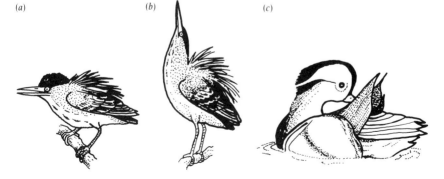

courtship. For most of the year, animals tend to keep their distance, avoiding getting near enough to others to risk being pecked or bitten. But you cannot mate that way! When it comes to courtship there is always the risk that the prospective partner may not be ready, so that a male about to mount a female is best to keep his options open and be all set for a quick retreat if she should attack.

This may explain why courting animals show intention movements, but then why should such actions have turned into displays? To do so they would have to signal something useful, that it paid the animal producing them to inform its partner. Here the answer may be that the male is simply telling the female: 'I'm off if you don't do something about it!' If mates are hard to come by it may be worth the female becoming receptive rather than risking the loss of the male.

2. Displacement activities. We have already discussed these in Chapter 4: actions which appear out of context to the observer, like grooming in the middle of courtship or nest building in the midst of a fight. They often occur when other actions are thwarted, as when a male is trying to court an unreceptive female, or when the animal is in a conflict. As mentioned above the conflict between approach to mate and avoidance of being injured may often occur in animal courtship. The occurrence of grooming and bathing movements during courtship are especially common amongst ducks, the most famous example being in the mandarin duck, in which the male has a greatly enlarged secondary wing feather over which he draws his beak when displaying (see Figure 6.10c). This action is clearly derived from preening, though the feather is not actually groomed during the display. The structure has also become greatly enlarged to show off the movement to best advantage.

Again we can ask what the gain is to the displaying bird in grooming movements having become incorporated into displays. It is not easy to be sure about this but perhaps, again, the male indicating a conflict about whether to approach and mate or to fly away might originally have spurred the female on to becoming receptive lest she lost her partner.

3. Thermoregulatory movements. Animals that indulge in strenuous activity, like courtship or fighting, tend to overheat, and various reactions help them to cool down. A coyote strutting stiff-leggedly around its kill will show its teeth and raise the fur along the back of its neck to any others that appear (Figure 6.11). Showing the teeth obviously displays its weaponry: a show of aggression which hardly needs a complex explanation. Raising the fur helps air to circulate close to the skin and its original

Figure 6.11. When coyotes threaten each other they strut around with the fur on the back of their neck raised. This is probably a cooling response adapted to serve a display function (photograph by F. J. Camenzind.)

function may have been to assist the animal to cool down in readiness for fighting. It is a very striking sight, however, and there is little doubt that it acts as a signal to others to keep away as well as helping the animal to 'keep cool'.

4. Protective movements. Just as aggressive displays often involve showing off the armoury of antlers, horns or teeth that a rival risks encountering, so the animal about to get into a fight will benefit from hiding its more vulnerable parts. It may avert its eyes, or frown so that they are better protected; it may flatten its ears making them less likely to be torn. All these actions may occur in the heat of a fight, but they also appear when animals threaten each other, which suggests that they have become displays indicating that an animal is prepared for a fight.

All these aspects of behaviour provide fertile ground for the evolution of displays, because they indicate something about the state of the animal that shows them. This information in turn may affect the way its partner or its rival behaves towards it. Provided this influence is beneficial to the performer then its behaviour is likely to evolve into a display. Just how then do displays come to differ from the behaviour patterns from which they are derived?

A major difference occurs through the influence of a process which Julian Huxley called ritualisation. If we think about ducks again, when they are bathing they splash up and down, throwing water over their backs, and they beat their wings to wet them thoroughly. The amount of each action, its intensity and the sequence in which the different movements are shown vary enormously. Likewise in grooming, ducks will nibble at the edges of each feather in turn, paying sufficient attention to each to get it back into shape and approaching it from outside or inside the wing presumably according to which best achieves its ends. The equivalent movements in display are quite different, being stylised and stereotyped, the actions often exaggerated but usually very constant in form and in length. The head toss of the drake goldeneye, referred to in Chapter 2, may well derive from a bathing movement, but it is very fixed in form and in duration, it is exaggerated in that the head is thrown right over till it touches the back and it is stylised in that it does not serve even to wet the plumage. Similar features can be seen in the mandarin's wing preening display. It is rather fixed and constant in form, it is always directed at one particular feather and it does not serve to clean this in any way.

Ritualisation has served to make displays clear and unambiguous so that the observer is seldom in doubt as to whether the animal is displaying or not (Figure 6.12). Many displays are remarkably striking and can be seen and heard from long distances away. They have usually ceased to serve the function of the behaviour from which they were derived. Niko Tinbergen referred to them as being emancipated, suggesting that their motivation had also changed from the original. He argued that the preening movement of a courting duck is part of its sexual behaviour system rather than that controlling grooming. It is thus more likely to be influenced by sex hormones than by the amount of dirt on its feathers. Indeed evolution may have gone one stage further than this and led displays derived from intention movements or displacement activities to occur when there is no intention, no conflict and no thwarting, even if these were essential for the display first to appear.

Ritualised behaviour patterns are often very constant in form. Although people can show a 'slight smile' or a 'broad smile', many animal displays are not graded in this way. A cricket chirps or it does not chirp, a fighting fish raises its gill covers in threat or it does not do so. Displays tend not to be half-hearted, but to be all or nothing, appearing at full throttle or not at all. Desmond Morris referred to them as showing 'typical intensity' because of this and he argued that they did so because it avoided any possible ambiguity. Thus, when an animal shows a display it is absolutely clear that it has done so and also which of its possible displays was used.

John Maynard Smith has pointed out that typical intensity may also have arisen for a very different reason, and this is because it is beneficial to animals to keep their cards rather close to their chests, especially when contesting with each other for resources such as territory. The animal that reveals its true motivation ('This little bit of ground would be quite useful to me but I don't really need it') is giving valuable information to its rival, and the rival can always make a higher bid by bluff: 'I demand this piece of land and am prepared to fight to the death to get it'. Thus negotiations

Figure 6.12. Julian Huxley carried out a pioneering study of displays in the great crested grebe during which he developed the idea of ritualisation. Four of the most striking displays are illustrated here: (*a*) The cat display in which the white wing bars are shown off; (*b*) the Dundreary attitude, so named after the long side whiskers of T. Taylor's fictional character of that name; (*c*) the shaking display, where the members of a pair face each other in a very erect posture and shake their beaks from side to side; (*c*) the weed dance, in which the partners both surface from diving with weed in their beaks and then swim towards each other, rising up and treading water as they come together. (After J. S. Huxley (1914), *Proc. Zool. Soc. Lond.* **35**, 491–562.)

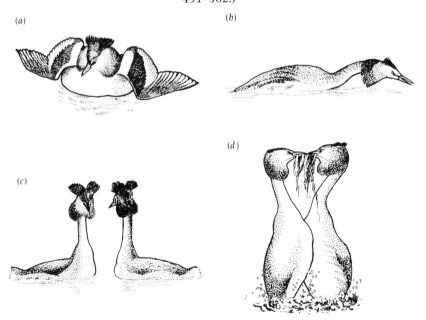

(*a*) (*b*)

(*c*) (*d*)

such as these tend to have become highly exaggerated and fixed in form with each side demanding the maximum. There is an amusing parallel here with trade union negotiators threatening all out strikes unless given vast sums, while managers offer minute rises with the excuse of imminent bankruptcy. Though not, of course, for evolutionary reasons, these demands and offers show typical intensity and are a far cry from the actual moderate figure at which both parties are really prepared to settle!

Ethologists have always been especially interested in courtship, so that the origin and evolution of displays is a subject on which they have generated many ideas. These ideas are plausible, but the problem of looking back at behaviour in evolutionary time remains. We will return to considering displays in Chapter 8, for they are the stuff of communication. Meanwhile we will consider the adaptive significance of behaviour: just how well has evolution honed it to the constraints of the present-day world? This is another aspect of evolution, but it is one that it is easier to study here and now.

7

Function

The word 'function' has many different meanings. Mathematicians manipulate functions, politicians attend them, and students follow courses on how computers function. But to biologists the word has taken on a very specific and precise meaning to which they restrict its use. The function of a feature such as a behaviour pattern is its selective advantage or survival value, the reason why individuals are thought to do better as a result of possessing it. This may seem straightforward enough, but it is a rather easy topic on which to think up ideas (as many popular books about animal behaviour make clear!) and yet a rather difficult one on which to test these ideas and so decide between them. Nevertheless, the study of function has become a popular and exciting field within ethology during the past two decades, and a good deal of useful information has accumulated about it. Before describing some of these results, we must consider the concept of function itself in more detail. At the end of the chapter we shall consider some of the recent changes in thinking about how evolution works that have led the ideas surrounding this concept to move to the centre of the ethological stage. These are mainly concerned with social relations between animals, so this will form a basis for the topics of the last two chapters.

7.1 The concept of function

The function idea can be best illustrated by taking a particular example. Let us do this by using the question 'Why do birds sing?' If you asked several people this, you would be likely to get very different replies. Some might say that it was because of sunny weather or because they are male and have the appropriate sex hormones circulating in their blood. These are causal explanations, of the sort we discussed in Chapter 4, rather than the functional ones that concern us here. Other people might say that birds sing to attract a mate or to drive away rivals. These are advantages that song may have and so they are possible functional explanations, based on the consequences of the behaviour rather than upon what causes it. These two sorts of answer are not alternatives in any way. All behaviour has both causes and functions, and understanding it fully involves knowing about both. But the two sorts of questions must be kept separate, for one certainly cannot answer causal questions with a functional explanation or vice versa, though people commonly try!

Some functions are easy to understand, without the need for a lot of argument or for complicated experiments. The functional significance of escaping from predators or producing eggs is obvious, for survival and reproduction are clearly of selective advantage, and few would question actions which lead to them directly. But there are some actions which are much more difficult to explain in these terms. Perhaps bird song does exclude rivals, but what is the function of that? Less competition over food maybe and hence more food available for the animal to feed to its young and hence the possibility of producing more young. Producing more young is without doubt advantageous in most species, as selection generally favours the individual leaving the most surviving offspring. But it takes several links in a chain of argument to go from song to this direct advantage, and other possible lines might be followed instead (Figure 7.1). Only by careful study can one decide which of them is likely to be correct.

Figure 7.1. Understanding the functional significance of a behaviour pattern involves discovering how it increases the transmission of an individual's genes into the next generation. This is the 'ultimate' function of behaviour but, in a case such as song, it cannot be simply linked to the act itself. As the diagram shows, song may have several immediate consequences each of which might lead to greater breeding success. Which of the links of argument is true in a particular case is a matter for study.

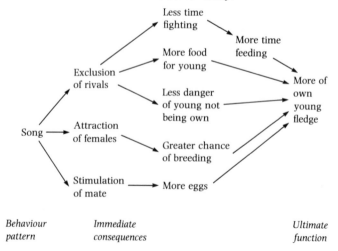

Behaviour   Immediate   Ultimate
pattern   consequences   function

An important point here is that behaviour patterns do not necessarily have only one function. Indeed any consequences that they have which lead their possessors to be more successful may help to maintain them in the population. Bird song does exclude rivals, as has been shown by the fact that birds made unable to sing suffer more intrusions on their territories. Song is also known to attract females. In several species, females have been found to approach a loud-speaker playing the male's song, and even to flutter their tails up and down in the posture of receptivity normally

Figure 7.2. Communicating may have both advantages and disadvantages, and the former must outweigh the latter if it is to be favoured by selection. In the example shown here, a male great tit gains if his song attracts a mate (*top left*) or repels a rival (*bottom right*), there is no cost or benefit to it being heard by most other species (*bottom left*), but a disadvantage of singing may be that it attracts predators (*top right*).

shown to a male. Finally, a third function of song is to stimulate the reproductive system of females: the ovaries of canaries and budgerigars grow when the birds are played tapes of their own species song, and they will eventually lay eggs without a male being present.

Song, therefore, is known to repel rivals, to attract mates and to stimulate them once attracted. In any one case all these three advantages may be important. But song may also have quite a few disadvantages. A bird singing loudly and obviously on top of a tree must be a sitting target for a hawk; some birds produce literally thousands of songs a day during the breeding season, so they are also using up a lot of time that they could be spending feeding or incubating their eggs. The important point is that, for a behaviour pattern like a song to be advantageous, the various benefits it confers must outweigh the various costs (Figure 7.2). The result is often a compromise.

The study of function often comes down to looking at behaviour in this way, as if it were a branch of economics, trying to discover just what are all its plus and minus points and, when they are added together, why the answer is positive. But there are several different approaches to examining functional questions: the next section will describe some of them.

7.2 Experiments and observations

A problem with studying the adaptive significance of behaviour is that it has evolved to function in a particular environment and so it can only be fully understood in relation to that environment. This raises difficulties, though it does of course give ethologists a good reason to work in some magnificent places! The main snag is that the usual approach of science, the carrying out of carefully controlled experiments, may be out of the question. There are many features of the outside world, like the weather for example, that cannot be changed at the whim of an experimenter and which may also alter from day to day. Even if the scientist can change a feature and so study the effects of this, it is not often easy to make the change simple and precise so that a comparison can be made between experimental and control groups which differ only in this one very specific way.

As we shall see, some excellent experiments have been done in this area, but much information has also had to come from observation alone where the experimental approach has been impossible. Observations are most fruitful where they also involve comparisons, not in this case between experimental animals and controls but between different species or between how one species behaves and how theory predicts that it would be best to behave. We will discuss these three approaches in turn.

7.2.1 Experiments

The classic example of an experiment on function was carried out by Niko Tinbergen and his colleagues on egg-shell removal in black-headed gulls (Figure 7.3). Shortly after a chick hatches, these birds will remove the broken shell from their nest and carry it away. Tinbergen asked what the function of this behaviour might be. Perhaps the chicks were sometimes cut by the broken edge of the shell if it was left; maybe fluid remaining inside the shell might be a breeding ground for bacteria which could infect them; perhaps predators might be attracted to the chicks and unhatched eggs by the conspicuous white inside of the broken shell. To test this last idea Tinbergen laid out a number of artificial nests containing gulls' eggs, some with broken shells beside them and some not. Returning later he found that the presence of the shell had indeed led to greatly increased egg loss to predators, which could only have been because these were attracted by the broken shell. This suggests that avoidance of predation is one reason why gulls remove empty egg-shells, though this need not be the only reason. He did not test them, so the other ideas Tinbergen had may also be correct.

An interesting point about this behaviour is that the gulls do not normally remove the shell until about an hour after the chick hatches, but

Figure 7.3. An adult black-headed gull removes the broken egg-shell shortly after one of its chicks has hatched. Experiments by Tinbergen and his colleagues showed that such broken shells attract predators and thus suggest that removal is advantageous because it enhances survival of the young.

remain with it during this time while its down dries out. The reason for this appears to be that unguarded chicks are often eaten by other gulls in the colony and they 'slip down' the throat best in the period after hatching when they are still wet. To leave the chick unguarded at this stage would thus risk predation from another source.

Figure 7.4. An experimental demonstration that song acts to repel rival males in the great tit. All the males were removed from the eight territories shown in (*a*). Three were then occupied by loud-speakers playing great tit song, two by loud-speakers producing a similar phrase on a tin whistle, and three were left silent (*b*). Males re-occupied the wood as shown in *c*); the part where song was being played was occupied more slowly than the control areas. (Redrawn from J. R. Krebs (1977), in *Evolutionary Ecology*, ed. B. Stonehouse & C. Perrins, pp. 47–62, Macmillan, London)

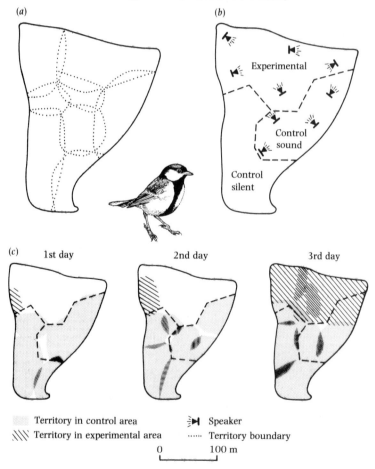

(*a*)

(*b*)

Experimental

Control sound

Control silent

(*c*)   1st day   2nd day   3rd day

Territory in control area   Speaker
Territory in experimental area   ⋯⋯ Territory boundary

0   100 m

We can return to bird song for another good example of an experiment designed to test a functional idea. John Krebs removed all the breeding male great tits from a small wood in which there were eight territories (Figure 7.4). He left some parts of the wood empty, but stocked others with loudspeakers, some of which played tapes of a tune on a tin whistle, while others produced recordings of great tit song. Observation over the next few days revealed that new great tits gradually moved into the wood and set up their territories, but they did not occupy all areas at the same time. The silent part of the wood and that where the whistle tune was relayed were taken over more quickly than the part where songs were being broadcast. This showed that song does indeed act as a 'keep out' signal to male great tits, so that one of its functions is to advertise ownership and so repel rivals.

7.2.2 Comparative evidence

The experiments described above were neat and allow firm conclusions. While results have less certainty when experiments are impossible they can, nevertheless, provide some convincing pointers when differences between species are studied in detail.

Niko Tinbergen's student, Esther Cullen, did the work for her doctorate perched on top of a cliff watching kittiwakes, small gulls that nest in dense colonies on cliff ledges in the higher latitudes of the Northern hemisphere. This nesting place is very different from the open areas where most other gulls breed (Figure 7.5) and the kittiwake's behaviour also differs from that of other gulls in many ways. Cullen concluded that the differences could be traced back to consequences of the cliff-nesting habit.

The number of ledges on cliffs is limited and, in keeping with this, kittiwakes guard their nest sites jealously, forming their pairs there well before egg-laying and doing all their courting and mating on the nest. Other gulls nest in open areas where sites are numerous and so they only need move to them just before laying as they are not in short supply. The kittiwake's nesting place is also much more dangerous than that of other gulls and, doubtless because of this, their nest building is more elaborate. They enlarge the ledge with mud and vegetation and scrape out a deep nest cup. Not surprisingly for animals living in such a precarious position, both adults and chicks have short legs and strong claws and, during copulation, the female sits rather than stands, as do other species. The young ones flap their wings very little when fledging approaches and, even if attacked, they will not leave the nest. With a fall of hundreds of metres just beside them, few would doubt that this is functional! Another

Figure 7.5. A colony of kittiwakes at their nests on a cliff face. Note various ways in which this species differs from ground nesting gulls: the legs are short so that balance is easier, the ledge has been built up with mud thus making it safer, the chick is conspicuously patterned and its droppings are allowed to accumulate around the rim of the nest. These last two features are thought to have resulted from lack of predation pressure on birds nesting on cliffs (photograph ©
J. C. Coulson.)

difference from other gulls probably also arises because the nests are on separate ledges and the young cannot move about between them: adult kittiwakes do not recognise their own chicks, presumably because they simply do not need to.

A further set of features that Esther Cullen identified as related to cliff-nesting, she attributed to the fact that few predators can get at nests on cliffs so, while dangerous in one way, these are safe in another. As one would predict from this, kittiwakes seldom produce an alarm call. The nests are also not camouflaged in any way like those of other gulls: thus the droppings of the young are not removed and the chicks themselves are far from cryptic, being strikingly patterned in grey, black and white.

Cullen's analysis of the relationships between all these different features is a convincing one, though certainty is not possible as all she could do is point to correlations between them and the fact that kittiwakes nest on cliffs. Her conclusions are the more firm, however, because some other gull species which are not closely related to the kittiwake, share the cliff-nesting habit and have the characteristics that go with it. The Galapagos swallow-tailed gull is an example here.

Another contrast between closely related species came to light from a study by Tim Clutton-Brock of two species of monkeys in East Africa (Figure 7.6). The black and white colobus and the red colobus overlap in their ranges and both live in forests where their diet consists primarily of foliage. In many other ways, however, they are very different. The red colobus is restricted to wet forests where it lives in troops of 40 or more, ranging over about a square kilometre. They feed on the flowers, shoots, fruit and leaves of many different types of trees. The black and white, by contrast, lives only in dry forests where it feeds on the mature leaves of a few tree species. Its troop size is usually less than ten and its range much more restricted than that of the red.

What is likely to have led to these differences? Clutton-Brock argued that the primary factor may be that the red colobus requires a more varied diet. If this is so it can only live in wet forests as it needs an area where some trees are in fruit at all times of year. This requirement will also constrain it to feed on many species and to have a large range so that food will always be available. Having a bigger group enables efficient usage of resources in a range which is large and helps in its defence.

This is a plausible argument and again, as with the kittiwake, it suggests that substantial differences in the behaviour of related species may arise for rather slight reasons. It is not, however, so easy in this case to be sure about cause and effect. There is a lot of argument involved in the claim

Figure 7.6. (*a*) The red colobus (photograph © T. H. Clutton-Brock) and (*b*) the black and white colobus (photograph by D. J. Chivers): two species of monkey which have been found to differ considerably in their group sizes and home range sizes, probably as a result of their very different feeding requirements.

that the diet is the main reason for the differences between the species and some of the threads of this could run in quite different directions. For example, a larger home range may make it possible to eat fruit at all times of year rather than being a result of the need to do so. All that is known is that certain features are correlated, but it is not easy in cases such as this to carry out experiments to discover which causes which.

7.2.3 Optimal animals

Comparing how an animal behaves with how theory predicts that it ought to behave is yet another approach to studying the function of behaviour. Given that natural selection favours those individuals that leave the most surviving offspring, if we examine the reproductive behaviour of animals, is there reason to believe that they do indeed do this? At a more day-to-day

Figure 7.7. (*a*) Individual bumble-bees specialise on particular flowers and so can develop the skills needed to exploit the nectar. Specialists on gentian blossoms, for example, must learn to pry apart the corolla tube and crawl into the base of the flower. (*b*) The route of an individual bee around its feeding territory, showing that it concentrates almost entirely on flowers of one species (in this case *Aster*, clumps of which are cross-hatched) and covers the area systematically so that the blooms have time to replenish their supply before it revisits them. The two different routes shown are for foraging trips six days apart. (*c*) Flowers of different species vary enormously in their sugar production (*above*), but fewer bees feed on those in which this is low so that many of the flowers end up offering roughly similar food rewards (*below*). (After B. Heinrich (1979), *Bumblebee Economics*, Harvard University Press, Harvard.) For (*b*) and (*c*), see pp. 130 and 131.

(*a*)

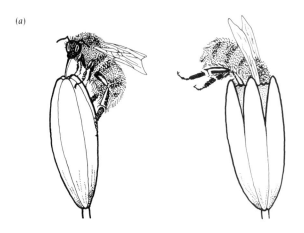

130

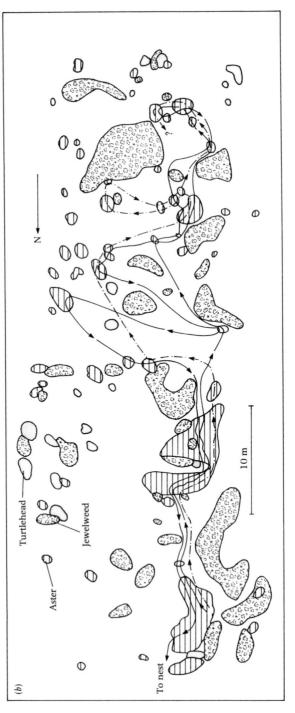

Turtlehead

Jewelweed

Aster

N

10 m

To nest

(b)

Figure 7.7b

level, many studies have been carried out on the efficiency with which animals feed to see whether this is as great as it could be. The usual assumption here is that the 'optimal forager' should spend as little time and energy feeding for the greatest possible returns or, in other words, that it should 'maximise net food intake per unit time'. Generally this

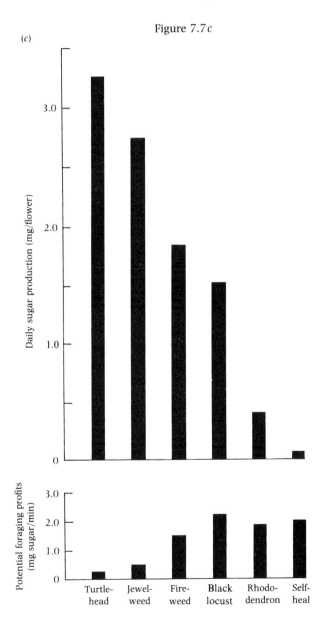

Figure 7.7c

assumption holds up rather well when the feeding of animals is studied in detail.

The relationship between flowers and the insects that pollinate them is a close and fascinating one in this respect. Insects fly from flower to flower feeding on the nectar which is produced for them as a reward for carrying pollen from one flower to the next. But this is only worth while to the flower if the insect takes the pollen to one of its own species, for pollen is the plant's equivalent of sperm and is wasted unless it fertilises an egg of the same species. Insects, however, often specialise on the flowers of one species because skills are usually necessary to reach the nectar and, having developed these, they are better equipped to do so. Indeed flower species probably differ from each other in shape very substantially for this very reason: only insects that have developed the appropriate skills can feed on them, and this forces the individual insect to stick to the same species and thus carry pollen from one plant to another of the same sort.

Bumble-bees are no exception to this rule, as has been shown beautifully by work on them by Bernd Heinrich in the United States (Figure 7.7). Each bee specialises on one particular flower, referred to as its 'major', while taking occasional nectar meals from other species, its 'minors'. Worker bumble-bees are short lived and tend to stick to the same major, but queens live longer and may switch from one to another. Having minors probably enables them to check up on whether another flower has become more profitable than that on which they are feeding most of the time. As flowering seasons are short, the abundance of different sorts of flower changes through the year and thus bees are well advised to check that the one they are feeding on still gives the best returns.

Whether it is worth a bee feeding on a particular flower species depends on several factors. Some flowers are much more common than others, so there is little flying time between one meal and the next. Some flowers also produce much more nectar than others so that the meals they provide are larger. Finally, the profitability of a type of flower depends on how many bees are feeding on it. If lots are doing so, then a bee looking for a food source will find that flower rather unprofitable, for there will be little nectar available on each visit. As a result, it is only common flowers with a high rate of nectar production that have large numbers of bees feeding on them. Each bee arrives shortly after another one and so receives little reward, but then it does not have to fly far to the next meal. On the other hand, few bees specialise on rare flowers with a low nectar yield and so each of them gets a large meal at each visit, though it has to fly a long way between one flower and the next to get this reward. Taking all these points into account, Heinrich argued that every bee does about as well as every other in terms of nectar intake per unit time. It follows that it would not pay any of them to switch to a different flower from the one they are majoring on.

A final point about foraging bumble-bees is that they tend to move round the area in which they live in a systematic fashion, visiting each clump of flowers in turn rather than going back to one they have fed on recently. This enables the flower to replenish its nectar supply and means that the insect gets a reasonable reward for each visit.

Bumble-bees, therefore, seem to be faring as well as they possibly could in the business of maximising the amount of energy they take in for the effort they spend feeding. Many other species have been found to do this too. Small birds tend to move systematically around their territories, allowing time for food to be replaced. If they find an item they will stay and search close by, for food items are often clumped, but they will move on when some time has elapsed without success. They will specialise, each individual tending to eat only those items which it is adept at handling. They are more likely to specialise in large and profitable prey than in small items which may take the same time to handle but which will yield much less reward. All in all, therefore, there is reason to believe that the foraging of birds, like that of bumble-bees, is organised in as profitable a fashion as possible.

That animals should have been found to forage in as efficient a way as they could is, in some ways, rather surprising. There are several reasons why we might expect them to fall short in this respect. For example, energy intake may not be what matters to them, for there may be all sorts of other factors, from the need to find minor nutrients to the necessity of watching out for predators, that will influence the way they behave. Another important influence may be whether or not we are observing them in the situation to which evolution has adapted them. In the laboratory we certainly would not be doing so but, even in the wild, the environment has been so modified by human activities that one should not always expect to find animals which are beautifully adapted to its present state. But studies of foraging behaviour do certainly show an impressive capacity on the part of animals to organise their behaviour in the most profitable fashion.

Throughout this section various ideas about what natural selection has led animals to do have been mentioned. In talking about foraging we assume that efficient animals do better than inefficient ones and that selection will favour the former. Earlier we discussed other ways in which animals enhance their own survival and reproductive success, on the assumption that this was to their advantage. Now, however, it is time to look at these assumptions a little more closely, for there has been something of a ferment in this area over the past two decades and this has revealed subtleties about the way that evolution acts which have strong implications for the way we view behaviour, and especially the relationships between animals of the same species.

7.3 The sociobiological revolution

The enormous current interest in the function and evolution of social behaviour stems from a synthesis between ethology and evolutionary theory which had curious and rather inauspicious origins some 25 years ago. Something of this history will help to put in context the growth of this subject, which is often now referred to as behavioural ecology or as sociobiology.

In 1962, V. C. Wynne-Edwards produced an impressive book called *Animal Dispersion in Relation to Social Behaviour*, in which he suggested that the behaviour of animals can be best understood if they are viewed as acting for the good of the species or group to which they belong. He proposed that many of the communal displays shown by animals, such as mosquito swarms, allow them to assess population density and so adjust the population size to match the food available. He also argued that animals do not fight very frequently because harming others would do damage to the group (Figure 7.8). On the other hand, they may share food with each other because this is to the group's advantage. If food is scarce, some of them may even forgo breeding so that the size of the group does not reach a level at which individuals would start to die of starvation.

Wynne-Edwards simply made clear the assumptions and thinking of many biologists at that time. The value of his contribution was not that it converted others, but that it caused them to look at these ideas in more detail and ultimately to reject them. If members of a group do not breed because food is running short and the group might suffer if they did, what would happen, some people asked, if a mutant animal arose which failed to obey this rule. The answer is simple: it would pass more of its disobeying genes to the next generation so that these would spread through the group as generations went by. Eventually all animals in the group would breed as much as they could, regardless of whether the group as a whole would benefit from restraint.

Points such as this led to acceptance of the idea that selection acts

Figure 7.8. A male Uganda kob courts a female while another waits nearby. In this species, each male occupies a small territory about 20 metres across and 10 or more of these territories are packed into a traditional breeding ground to which females come to be mated. There is a great deal of communal display and this is just the sort of phenomenon that Wynne-Edwards thought to be concerned with the assessment of population density. (photograph by P. A. Jewell.)

primarily on individual animals, favouring those that do better in the business of passing on their genes at the expense of those that do worse. The term 'fitness' is often used in this context to describe the capacity of different individuals to contribute to the next generation; those of greater fitness make a larger contribution so that their genes tend to spread. Such expressions as 'acting for the good of the species' have, as a result, dropped almost totally from use. This change in thinking has strong implications for behaviour, and especially for the social relationships between animals. How can one explain the huge number of examples that Wynne-Edwards amassed to support his argument, if the very basis of it is unsound?

If individuals act only for their own good, aggression between them becomes easier to understand; indeed one might expect them to fight a tremendous amount the whole time, each being out for its own ends and careless about possible damage to others. This certainly does not occur, but the reason is probably simply just that fighting is dangerous. Even if one animal is much bigger than another, the small one may give it a scratch that will go septic or a fatal nick in the jugular vein before it is subdued. Although the risk of injury may be very slight, it will not be worth taking unless the fight is over something very valuable. Nevertheless fights to the death have often been described amongst animals. When ethologists have carried out thorough field studies, for example watching a small troop of monkeys for a few thousand hours, it has not been unusual for one animal to be seen killing another. It would be quite exceptional to observe such a high rate of murder in a human society, despite the supposedly high aggressiveness of our species!

A good example here, and one which certainly would be hard to fit in with thinking in terms of the good of the species, is what occurs when male lions take over a pride. A pride consists of a small number of males, females and their cubs. Some males also go around in bachelor groups that wander over the savannah looking for prides to take over. This they are likely to succeed in doing if they come across a pride in which the adult males are few, aged or out of condition. There will be a scrap, at the end of which the previous males are driven off. One of the first things that the new males do thereafter is to kill the cubs already in the group, behaviour one might think of as wanton cruelty. It is, however, easy to see how it benefits the males themselves. They may only be able to control the pride for a short time and the more young they have during that period the more of their genes they will be passing on. By killing cubs that they did not father, they make the mothers stop lactating and become receptive and thus they speed on the time when these females will bear cubs sired by themselves. In this way infanticide, which also occurs in many other species, can be understood to be of advantage to the individual, though it would certainly not be to the group or the species.

Aggression in animal groups is, therefore, not very rare. It tends to be limited where it could have a dangerous outcome for either of the participants: in this case it is commoner for animals to display and threaten each other until one backs down, an aspect of communication to which we shall return in the next chapter. But what then of altruism? Surely selfish animals should not share bananas and yet chimpanzees often do this. The young of many bird species, from European long-tailed tits to Arabian babblers and Florida scrub jays, remain with their parents and help to feed later broods rather than going off to have families of their own (Figure 7.9). Most remarkably, some social insects like ants and honey-bees have workers that are completely sterile so that they never breed themselves. They have sacrificed all their own fertility and devote themselves selflessly to raising the offspring of others in the colony. Usually, as with helpers

Figure 7.9. Helpers at the nest: in many bird species other individuals help the parents to feed their chicks. In most such cases the helpers are the young of previous broods. This photograph shows a mated pair of red-throated bee-eaters and (in the middle) their son, hatched in the previous year and now helping at their nest (photograph by C. H. Fry.)

at the nest in birds, the young ones that they raise are their own brothers and sisters.

A major contribution to our knowledge of how evolution works, which helps us to see how such behaviour could have arisen, came in a theory put forward by William Hamilton in 1964. He took as his basis the point that what really matters as far as natural selection is concerned is the extent to which genes are passed on from one generation to the next. Genes leading animals to behave in ways that enhance this transmission will spread at the expense of others. Now animals producing more young than their rivals are clearly succeeding in this way. But what of those that help their parents, their sisters or their cousins? Relatives share genes, to a diminishing extent as the relationship gets weaker, so assisting close kin may actually help an animal to get more of its genes into the next generation, though in a more roundabout way than by breeding itself. This theory is referred to as that of 'kin selection' because it explains why animals are prepared to behave in ways that involve sacrifice on their own part but advantage to their kin. The argument put forward by Hamilton is that animals should maximise not just their own fitness but what he refers to as their 'inclusive fitness'. This has both a direct component from their own reproductive efforts and an indirect one from the help they give to relatives. A young bird which stays to help its parents with the next brood will therefore be making a contribution to its own inclusive fitness. If breeding requires skills, it may also be gaining useful practice and, if good territories are few and far between, it may have more success on the short term by staying at home than by going off and trying to find one of its own. In the long run it may even achieve 'promotion' by becoming a breeding adult itself once its older relatives have died. Though it appears at first sight to be selfless altruism, helping at the nest may therefore be very much in the bird's own interests and not, strictly speaking, altruism at all.

Aggression and altruism are just two aspects of behaviour that we can understand better when we start to think in terms of animals acting so as to maximise their inclusive fitness. But, for most animals, their major contribution to the next generation is through their own reproductive efforts. In jargon terms, they are maximising the direct component of their inclusive fitness. We will close this chapter by considering various ways in which this is achieved.

7.4 Strategies of reproduction

Animals which mate with one another are not usually close relatives, as inbreeding often leads to weaker offspring and so tends to be disadvantageous. Nevertheless, sexual partners have a shared interest in the young that they produce and so they often cooperate in raising them. Indeed one might think they should always do so to maximise the chances of the young surviving. But this is far from so. Many birds, and a few mammals such as marmosets, gibbons and some humans, are monogamous, forming stable mated pairs, the two members of which combine to rear their young. But there are several other ways of doing things. Toads, for example, produce hundreds of eggs and leave them to fend for themselves. In ducks, the males appear to have an easy time of it: they mate and then go off to moult in stag parties, while their females lay, incubate and raise the young unaided. Sticklebacks, as was mentioned in Chapter 1, are the opposite. The male builds the nest and tends the eggs and young, the female contributes only her eggs.

One can begin to understand why different species adopt different strategies, if one looks at just what may be the costs and benefits of each way of doing things. Various theories, some quite complicated and mathematical, have been devised with this end in mind, especially by Robert Trivers and others influenced by him. Here we can do little more than give some impression of the reasoning behind them. Young which taste nasty, like toad tadpoles, and which can fend for themselves as easily whether their parents are around or not, might as well be left. The parents will leave more descendants by putting all their energies into eggs and sperm and none into caring for them. But more vulnerable offspring have no chance of survival without parental care. Just how they are cared for depends on the advantages to each of the two parents. If both together can rear twice as many young as could one alone, or if one of them leaving its partner to tend the brood has little chance of finding another mate, then both are likely to help. On the other hand, if one parent can do quite well on its own, then it may pay its partner to desert. The one that deserts is, of course, behaving in a 'selfish' manner but, if it gets more genes into the next generation as a result, natural selection will favour it deserting.

Which partner is likely to be the one that deserts? The answer here seems to be that it will be the one most likely to be able to breed again easily. If there are a lot of unattached females around, a deserting male may well

Figure 7.10. In a wood occupied by pied flycatchers, a female attracted to a male cannot tell whether he is unattached, and so will help her rear her brood, or has another female incubating elsewhere, to which he will return, thus leaving her to look after her young alone. (Redrawn from B. Silverin (1983), in *Hormones and Behaviour in Higher Vertebrates*, ed. J. Balthazart, E. Pröve & R. Gilles, pp. 388–97, Springer-Verlag, Berlin.)

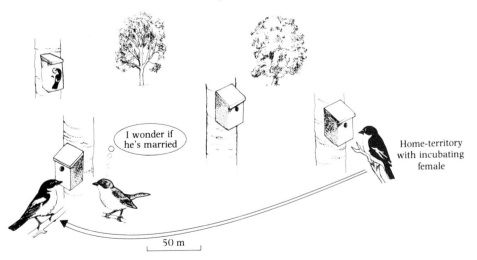

I wonder if he's married

Home-territory with incubating female

50 m

find a new partner and, provided he does not have to expend weeks of effort in getting another territory, defending it and building a nest before his second breeding effort pays off, it will be worth his while to have a series of mates rather than sticking to one. A good example here concerns the pied flycatcher, a small European songbird (Figure 7.10). These birds nest in holes in trees, or in nest boxes, the female incubating the eggs and both partners sharing in care of the young. In this circumstance they raise, on average, about five chicks. Sometimes, however, a male will move off while his female is sitting and occupy a new hole some territories away with a second female. After this female has started to incubate he then returns to help his first one care for their brood. The second female is therefore deserted and must rear her young alone; nevertheless she usually succeeds in fledging over three, so she does gain more than if she lacked a partner altogether. Furthermore she cannot avoid her situation, for there is no way a female pied flycatcher can tell whether a male that comes along has another mate elsewhere or not. If he has not she does better: the first female has the same number of young as she would do were her male not bigamous. The real gain, however, is to the male, who fathers eight or more surviving offspring, far more than he could achieve with only a single partner.

Females too may move from one partner to the next, though this is less common. Sticklebacks do it, possibly because there are abundant males or because after the huge effort of egg-laying the female would have little energy for parental care. She may thus gain more by leaving the male to care and, if she can manage to do so before the breeding season ends, feeding so as to build up a further supply of eggs that she can leave with a second male. Generally, however, desertion by the female is rare for several reasons. One is that eggs are a very large investment and it takes a lot of time and energy for the female to replenish her supply, while males can produce huge numbers of sperm at frequent intervals. They can therefore go from mate to mate fathering large numbers of offspring, while the gain in changing mates to a female is much less. A second reason is that females are often literally tied to their offspring. A female mammal can hardly leave her young to fend for themselves or be looked after by their father in the middle of pregnancy! Nor indeed can she do so soon afterwards for they continue to be dependent on her for milk. A third point is that females have more certainty that their young are their own than do males, for it is seldom possible for males to guard their mates assiduously enough to be sure that they have not mated with others, while a female can be certain that an egg she laid or a young one she produced

Figure 7.11. A male redwinged blackbird singing on his territory in a reedbed. If successful he may attract several females to nest there, but this is not all gain to him as there is a risk that some of their eggs may have been fertilised by others (photograph by W. A. Paff for the Cornell Laboratory of Ornithology.)

is her own. In other words, males suffer from 'paternity uncertainty'. In many species, this is a real influence, for mating with another's partner while he is off feeding will certainly do wonders for a male's inclusive fitness: he can then help his own partner to raise their shared offspring while his cuckolded neighbour raises another brood for him as well. Such 'extra-marital' matings have been observed in many species and are probably widespread. One experiment on male redwinged blackbirds in North America certainly suggests this is so (Figure 7.11). In this species, each male defends an area of marsh in which several females may nest. It was found that, when males were vasectomised so that they could not father chicks, many of the eggs on their territories were still fertile. This was even the case when all the males in a patch of marsh were sterilised in this way, indicating that itinerant males without territories must have been mating with the females. In this case, mating with a second male was certainly to the female's advantage and, if males are sometimes naturally infertile, it may normally be so too even when an experiment is not in progress!

In most species all individuals of one sex adopt a similar reproductive strategy, and thinking about costs and benefits may suggest to us why they follow this one rather than any of the other possibilities. In some cases it is clear from this that one breeding system is indeed the best for all members of a particular species so that no animal that happened to adopt another one could do better than if it conformed. In the terminology devised by John Maynard Smith, there is in this case a single 'evolutionarily stable strategy' or ESS which it benefits all individuals to follow. In other cases, however, a mixed ESS can occur and it pays some individuals to do one thing and others another. The foraging of bumble-bees is a good case here, though not from the point of view of reproduction: the flowers it is best for a bee to visit are those being least exploited by others, so the bees end up with different strategies. The reproductive behaviour of male green treefrogs provides another instance. These animals form assemblies beside ponds and set up calling choruses to which females are attracted. However, calling is dangerous and energy consuming, as well as making it hard to keep a look out for the females that approach. Some males avoid these disadvantages by adopting 'satellite' status instead of calling. They sit in wait near calling males and watch out for females, intercepting them as they arrive and often succeeding in mating with them before they reach the caller to whom they are attracted. Not all males can become satellites, however, because then the pond would be silent and no females would come, so the stable situation is a population consisting of a mixture of the two sorts of frog.

These examples, drawn from the breeding behaviour of animals, show just how powerfully modern evolutionary theory can shed light on various aspects of the behaviour of animals. Many of its facets, which appeared hard to account for only a few years ago, have come to seem obvious with the help of the new body of theory which has recently been developed. In looking at function, we have to think of how various actions have come to be favoured by natural selection. It is now clear that survival and reproduction are not even all there is to this matter. Indeed survival, even into old age, is of no consequence in itself. It is the passage of genes from one generation to the next that matters, and long-lived animals do not necessarily have greater success in this. This revolution in our understanding of how evolution works has led to extensive study of animal behaviour and a rethinking of many old examples in terms, not of how particular actions benefit the species nor, strictly speaking, how they may be of advantage to the individual's own reproductive success, but of the ways in which every animal behaves so as to maximise its inclusive fitness.

In the next two chapters we will return repeatedly to ideas such as this from modern evolution theory for, especially in communication and the social relationships between animals, these new theories have done a great deal to illuminate the reasons why animals behave in the ways that they do.

8

Communication

Many of the examples discussed throughout this book have involved communication between animals, for this has always been a subject of particular interest to ethologists. Indeed the theories they developed were in some ways biassed by this interest. Acts of communication do tend to be very fixed in form, to act as releasers for other actions, and to develop very similarly in all members of a species. Perhaps ideas such as that of the fixed action pattern or the innate releasing mechanism would have come less easily to mind if ethologists had been more interested in grooming or feeding behaviour than in aggressive or courtship displays. But then they had every reason to be enthusiastic about studying the latter, because the signals that animals produce in communication are often remarkably striking. Discovering just what they are saying to each other is one of the most fascinating fields of ethology.

Whenever we see an animal that is brightly coloured or boldly patterned, or one that is making a loud and obvious noise, we can be pretty certain that communication of some sort is taking place. Standing out from the background is dangerous and there must be some benefit to doing so which offsets the risks involved. Most often the communication is between members of the same species, and the risks come when predators cue in on this as a means of finding a meal. Sometimes, however, different species communicate with one another and here it is often prey animals that forsake their camouflage and communicate directly with predators that might eat them. These different sorts of relationship between predators and their prey make a useful starting point as they illustrate some basic principles of animal communication.

8.1 Crypsis and communication

Normally, animals are as cryptic and silent as they can be, and it is amazing the accuracy of camouflage that some of them have achieved (Figure 8.1). Selection quickly removes those that fail to comply: the slightest rustle on

144

a woodland floor is enough for a barn owl to home in with deadly accuracy on the mouse that made the sound.

As with other behaviour patterns, then, communication has costs as well as advantages. The animal that advertises for a mate could well end up being eaten instead. This is nowhere better illustrated than in a frog (*Physalaemus pustulosus*) from Panama in which the males call in choruses to attract mates. There is, however, a problem: fringe-lipped bats home in on the calls and can catch and feed on the frogs with great success. The frogs cannot obtain mates unless they call, so they must do so, but the intense selection pressure that the bats provide has made them minimise their risk. On all but the darkest nights, they spot approaching bats and, while they often ignore the smaller species, their chorus will shut down the minute a fringe-lipped one appears above their pond.

Figure 8.1. The patterning on most animals makes them merge with their background and some of them are remarkably well adapted in this respect: a bittern with its beak pointing upwards to match the reeds in which it stands (*a*); stick insects that mimic patches of lichen (*c*); mantises and other insects that have flattened, green, leaf-like areas on their bodies so that they blend with the foliage in which they sit (*b* and *d*). (*b* and *c* redrawn from M. Edmunds (1974), *Defence in Animals*, Longman, Harlow, Essex.)

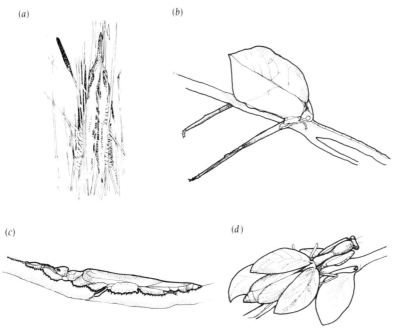

(*a*)   (*b*)

(*c*)   (*d*)

Figure 8.2. (*a*) A blackbird keeps very still and produces a 'seeet' call as a sparrow hawk passes over. (*b*) Sonagrams of these calls as produced by several species, showing how they are closely similar. (*b* after W. H. Thorpe (1961), *Bird Song*, Cambridge University Press, London.)

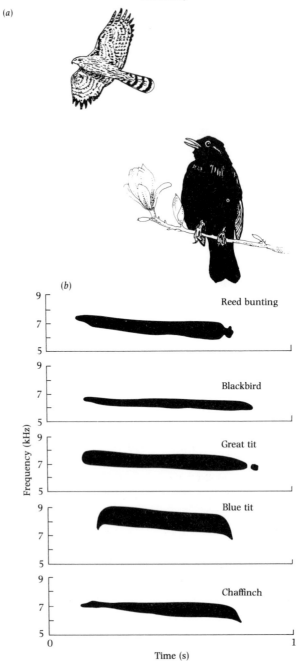

(*a*)

(*b*)

Another good example of communicating when a predator is around, at the same time as minimising the risk of being captured, is the high-pitched and thin 'seeet' alarm call that many small European songbirds produce when they spot a hawk (Figure 8.2). Interestingly, the call is almost identical in all the species that produce it. Two factors have probably contributed to this. First, there is no need for the call to incorporate information about which species the caller is, provided that its mate or young gather the information that there is a hawk about and so know that they should beware. Thus there is no reason for the calls to be different. Secondly, there is good reason for the calls to be the same. This particular sound has characteristics which make it especially difficult to locate, so all species seem to have homed in on the same form to reduce the risk of capture as much as possible. Animals usually locate sounds by detecting differences between their ears: for example, if a sound is louder in one ear than the other it obviously comes from the side of the head where it is most intense. However, with thin, high-pitched sounds, like the 'seeet' call, such differences are slight and locating the caller becomes very difficult for a predator. Certainly, to a human, the call is almost ventriloqual and it is very hard when hearing it to decide where the caller is perched.

These examples illustrate how pressure from predators may mould the form that animal signals take, assuming that most animals avoid communicating with predators as far as possible. But what about cases where signals are thought to be produced specifically to communicate with predators? Very often the information that these provide is misleading. Such signals are usually referred to as deceitful, without of course implying that the animals producing them are conscious of lying in any way: they make such signals simply because selection favours the transmission of false information. The plover that drags its wing along the ground as if injured and so distracts a fox from the area of its nest (Figure 8.3) is communicating with the fox, but the information it is providing is false. Its wing is not really broken and, once it has lured the fox far enough away for its chicks to be safe, it will fly strongly off. Likewise, the eyed hawk moth that flicks its forewings forwards to reveal two large eyespots deceives the bird about to eat it by mimicking a large animal with its eyes spaced wide apart. Insects which taste nasty or have stings often advertise their distastefulness by bright colouration so that predators soon learn that they are to be avoided. In this case both predator and prey gain: the prey escapes damage and the predator avoids an unpalatable meal. Other species have, however, evolved so as to mimic the unpleasant prey while being perfectly palatable themselves. The black and yellow stripes of wasps and bees warn birds of their stings; hoverflies are harmless but similarly striped and they may thus avoid capture without the need to be dangerous in any way (Figure 8.4).

Prey animals can thus warn predators to keep clear, or deceive them into doing so under false pretences. Predators can behave in a similar way, in this case appearing to be harmless when they are actually lethal. The polar bear is white so that it blends in with the snow and can creep up on its prey unseen. An American buzzard, the zone-tailed hawk, which preys on small mammals, also merges with its background, but it does so by being just like a vulture in silhouette. Prey animals pay little attention to vultures as they eat only carrion, so the hawk gains from this deception by getting

Figure 8.3. A ringed plover lures a fox away from her chick with the broken-wing distraction display. This signal benefits the bird but is to the disadvantage of the fox as it is deceived into moving away from an area where it would find an easy meal.

Figure 8.4. A case of mimicry. Both the hoverfly (*a*) and the wasp (*b*) shown here have black and yellow striped abdomens yet they are not at all closely related. The similarity probably evolved because the hoverfly, although harmless, benefits from the avoidance predators show towards it.

(*a*)   (*b*)

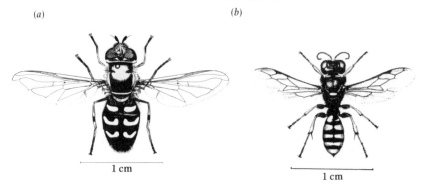

1 cm

1 cm

close enough to pounce without the prey running off. The green colour
and rather leaf-like limbs of praying mantises probably also help them to
sneak up on prey unseen. These examples are of predators being camouflaged. However, predators too may produce signals which mislead their
prey. An especially nasty example is that of the fire-fly *Photuris* (Figure 8.5).
Male fire-flies produce a signal of light flashes in a pattern typical of their
species, and females respond with another pattern which is also species-
specific, thereby enticing the male to approach and mate. The female
Photuris, however, is a predator, and when she detects the flashes made
by the male of another species she responds with the appropriate female
version. The unfortunate male flies in to mate and meets a sticky end
instead.

Figure 8.5. (*a*) The male *Photinus* firefly produces two flashes and a
female of the same species replies with her single more intense flash.
In (*b*) the two flashes are a human imitation of the pattern and the
reply is that of the female *Photuris* which attracts male *Photinus* in this
way and eats them. (Redrawn from J. E. Lloyd (1981), *Scient. Amer.*
July, **245**, 110–70.)

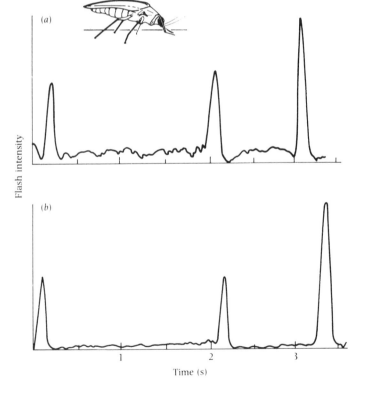

Time (s)

These examples illustrate the fact that predators and prey may either fit in with their backgrounds and so avoid communicating or may actually communicate with each other, though in this case the information they provide is often false. That communication does not always involve information which is useful to the animal receiving it is not true, however, just of predators and prey. Much of the communication between members of the same species can be interpreted in similar terms.

8.2 Information and manipulation

In the past, ethologists often used to refer to communication between animals of the same species as involving the sharing of information for their mutual benefit. Indeed, much communication may be just this. For example, when potential mates meet it is advantageous to both of them that they communicate to each other accurately information about the species to which they belong, for if they are not members of the same species both of them will waste their breeding effort. Strictly speaking, however, animals will produce signals when this is to their advantage, regardless of whether others benefit from receiving them. The examples in the last section showed this well, for prey animals send signals to predators when they gain by doing so and the gain they make is usually at the predator's expense.

Exactly the same rule applies within a species, as we might expect from the idea that animals are essentially selfish, behaving so as to maximise their own inclusive fitness. So, while the information that passes from one to another when a signal is transmitted may be useful to the one that receives it, this will not necessarily be the case, and there are some very good examples where it is better to think of animals as manipulating each other rather than sharing information. One of these is in the mating behaviour of the bluegill sunfish (Figure 8.6). Most males of this species do not mature until they are seven or eight years old, when they set up territories to which they attract females for mating. The females are smaller and look quite different. Sometimes two of them may spawn simultaneously with one of these males, sperm and eggs being shed into the water as they swim around together on his territory. Deceit enters into this scheme of things because some males, around 20%, mature at only 2–3 years old, when much smaller than the norm. These look just like females and join in the mating of spawning pairs. The territorial male does not drive such a rival off as he cannot tell that he is not a second mate, but actually he is a male and he shares in the fertilisation of the eggs that the female produces. Of course all males cannot mature early, or there would be no territorial ones to attract females in the first place, so this is an example of a mixed evolutionarily stable strategy, with a balance between the two possible modes of behaviour. In this example males mimic females and the other males, unable to see through the deception, are manipulated by the mimics. The case of the pied flycatcher mentioned in the last chapter can be put in very similar terms. Here the female cannot tell if a male is mated or not and so cannot choose an unmated one who will help her rear her chicks rather than the one who will leave her to return to his first mate.

Deception is possible in these cases because there are no cues that a signal receiver can use to discriminate between the classes of individuals involved, in other words to tell young male sunfish from females and mated flycatchers from unmated ones. In other cases, however, deceit is impossible because there are ways in which other animals can see through it. Some good examples here come from cases where it pays animals to appear to be larger than they really are. This is often so in aggressive displays for, the larger an animal appears to be, the more a rival is likely to be intimidated into retreating without a scrap. It is probably for this reason that Siamese fighting fish extend all their fins as far as possible when swimming alongside an opponent and raise their gill covers when facing one (Figure 8.7). Both actions make them as large and striking as they can be. If only some fish displayed like this, as was probably originally the case, they would be likely to have deceived others of the same size into retreat.

Figure 8.6. Three bluegill sunfishes spawning. The large animal at the back is a full-grown adult male. Though they look similar, only the nearer of the two small individuals is a female. That in the middle is a young male which is a female mimic. By this deception these fish avoid being chased away by large males and are able to steal some fertilisations from them. (Redrawn from M. Gross (1982), Z. *Tierpsychol.* **60**, 1–26.)

Thus appearing bigger would have been to their advantage and natural selection would have made it spread through the population until, eventually, all fighting fish looked as large as they possibly could when displaying. At this stage, however, the trick breaks down for each just looks an exaggerated version of its own size. If every fish can increase its apparent size by 20%, all will look that much bigger, but their sizes relative to each other will remain the same. Their opponents will gain accurate information about their size rather than being misled in any way.

Another nice example is in the calling of male toads studied by Nick Davies and Tim Halliday (Figure 8.8). Male toads cling onto ripe females in a posture called amplexus which ensures that they fertilise the eggs as they emerge. Males often fight to take up this position and the male already in amplexus kicks and croaks to repel boarders. Larger males generally win in such encounters, as one might expect from their greater strength. Being larger, however, they are also able to produce deeper sounds, and Davies and Halliday showed that these too had an effect. They silenced males in amplexus by fitting elastic bands round their arms and through their

Figure 8.7. The male Siamese fighting fish is a fine example of an animal that looks larger when displaying (*c* and *d*) than it does normally (*a* and *b*). The displaying fish raises all its fins when broadside to its rival (*c*) and its gill covers when facing it (*d*), thus making itself appear as large as possible. (Redrawn from M. J. A. Simpson (1968), *Anim. Behav. Monogr.* **1**, 1–73.)

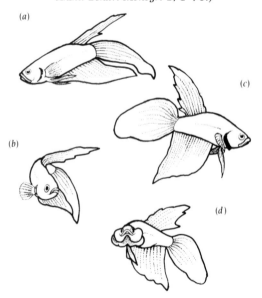

mouths. They then saw whether other males attempted to displace them. As expected, medium-sized males spent more time attempting to displace small males than large ones. However, if the deep croaks of a large male were played from a loudspeaker, a small male in amplexus was harassed much less by his larger rival. Clearly toads can use the calls to assess the size of others and will be less likely to fight with those whose calls are deeper. Here again then, the information transmitted is accurate and the animal receiving it acts on it appropriately. But why do small toads not cheat, developing deeper voices and so appearing larger than they really are? The answer is probably that the larger an animal the deeper the sound it is capable of producing. So, just as with the size of fighting fish, if all toads call as deeply as they can, the larger ones will still call more deeply

Figure 8.8. Male common toads try to displace each other from the amplexus position on the backs of females. Medium-sized males attack more often and for longer when the defender is small than when it is large (the four bars on the left are larger than those on the right). In these experiments the defenders were silenced by passing elastic bands through their mouths and round their forelimbs, and other male calls were played from loud-speakers. No matter what the size of the defender, the attacker would be more persistent if the call played was that of a small toad than if it was that of a large one (the left bar of each pair is larger than the right one). (Redrawn from N. B. Davies & T. R. Halliday (1978), *Nature* **274**, 683–5.)

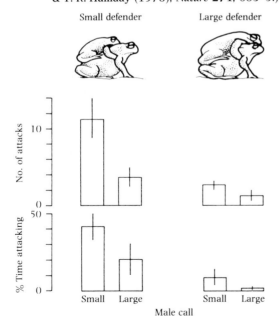

than the smaller ones, and depth of croak will be an accurate reflection of size.

To summarise, communication involves the transmission of information, in the form of a signal, from one individual to another. As we have discovered, this 'information' may be true or false, the important point being that producing the signal should be of benefit to the animal that does so. Sometimes deception may be possible, and the signaller may gain from it, but this will not be possible if the animal receiving the signal can detect that the signal is misleading. Deception tends, therefore, to be rather rare, for the more animals practice it, the more it will be best for the recipient to assume that the information it is receiving is false. If almost every plover which appears to have a broken wing is really feigning injury, then a fox discovering one would gain most by not following the adult but searching the area for a nest or chicks.

8.3 Messages and their meanings

A useful way of analysing communication was devised by the American ethologist W. John Smith. He pointed out that it could be looked at either from the point of view of the animal that sent the signal, in which case one would ask what the *message* encoded in it was, or from the viewpoint of recipients, to see what its *meaning* was for them.

Let us return to our singing bird. What is the message incorporated in the song and what meanings may it have for those that are listening? The most obvious point is that, as every bird-watcher knows, song is usually clearly distinct between species, so that it includes information as to the species to which the singer belongs. Furthermore, song tends only to be produced by male birds in breeding condition, they usually only sing when they have a territory and in some cases they stop singing as soon as they obtain a mate. In such a species the basic message of song may be 'I am an unmated adult male sedge warbler on my territory and in breeding condition'. In addition, there may be many subtleties incorporated in the message. Signals often vary from place to place, forming dialects like those in human language. Individual animals may also have idiosyncrasies in the form of the signal that they employ. They may have a range of different signals which convey subtly different information. The singing bird may thus also be saying where he comes from, exactly who he is and perhaps even something about his motivation if the songs differ according to whether they are directed at rivals or at potential mates.

Many animal signals are, like song, primarily concerned with reproduction, attracting mates and repelling rivals being their main role in communication. However, especially in species that live in social groups, there are many other messages that it may be useful to convey. Animals that move around in groups often have obvious patterns on them which can be seen some distance off and enable them to maintain contact with their companions. Call notes can serve a similar function, especially where visibility is poor. These are simple sounds and patterns, and they tend not to vary much. Their messages may simply be 'I am here', though usually they also communicate the signaller's species, and there are cases where, despite their simplicity, they vary enough to indicate individual identity.

The alarm calls referred to earlier are an example of another type of message, individuals in this case warning their companions of danger. The 'seeet' call of small birds is most reliably produced by the appearance of a hawk but, in the breeding season, birds with young may also call in this way when alarmed by a dog or a human near to their nest. Most animal signals are like this: the message they convey is a generalised one, rather than being precise like a word in our language. A famous exception is the alarm calling of vervet monkeys (Figure 8.9). These animals have several different calls which they produce when alarmed and three of these are specific to particular sorts of predator: a snake call, a leopard call and an eagle call. Here the information in the call is very precise: the calling animal might as well be shouting 'snake', 'leopard' or 'eagle', provided that its listeners were able to recognise the meaning of those words as we can do.

What do its companions do when they hear one of these calls? In other words, what *meaning* does the call have for them. The answer is that this too is very precisely defined. Animals hearing the leopard call rush up into trees, they respond to the eagle call by running down to the ground and hiding in thickets, and the snake call leads them to approach and look down. We cannot tell if they have a mental image of the predator in question, but they certainly behave as if they were aware of the particular threat it poses.

To understand messages, we have to study the particular situation in which an animal produces a signal, in other words try to discover what causes it to do so. In the case of meanings it is the recipient of the signal that we must study: how does it respond when the signal is received? This response may vary from one animal to another. In the case of a contact call, for example, other members of the flock may approach so that the group keeps together, animals from other flocks may keep their distance, predators may be attracted with the prospect of a meal, and animals of other species are likely to ignore it as irrelevant to their interests. The same

Figure 8.9. The response of vervet monkeys to three different alarm calls. (*a*) That produced in the presence of a leopard leads them to run into trees, (*b*) the eagle call leads them to run down into the undergrowth and (*c*) the snake call causes them to look down and approach.

(*a*)

(*b*)

signal can thus have many different meanings, depending on the listener and on the exact context in which it is received. To stress a point made earlier: the critical factor is that, for a signal to be produced, the advantages to the signaller must outweigh the disadvantages. Otherwise, there is no reason why evolution would favour signalling.

There are thus many different messages that animal signals may convey, though it is unusual for these to be very precise. Most often they provide rather general information about the motivational state of the signaller, though their features can also incorporate more specific information, such as the species or identity of the caller. The number of distinct signals that a species uses is not often large. What makes the signalling system flexible and varied is that the same signal may mean different things to different recipients and the same animal may respond to it in different ways depending on the context in which it is received. Thus, with a comparatively small repertoire of rather stereotyped signals, animals can convey a wealth of information from one to another.

8.4 The form of signals

The message is what an animal encodes in a signal it sends; the meaning is what another makes of it. The signal is the physical form in which the transmission from one to the other takes place. Just as the word 'dog' has neither four legs nor a bark, so the signal is not necessarily related in any

Figure 8.9c

(c)

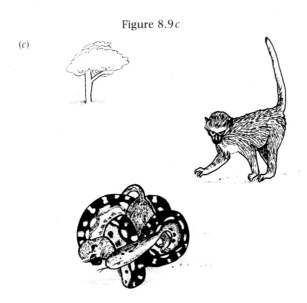

very clear way to the message that it conveys. However, certain types of
signals are better adapted to certain functions and it is illuminating to
consider the characteristics that make them so.

Most animal communication uses sights, sounds and smells, though at
close quarters the senses of touch and taste may come into play. Occasional
species may use a 'sixth sense', like for example the electric sense of some
fish: these can detect each other's discharges and so can signal using
changes in the patterning of their electricity production. Such bizarre
senses are rare, however, and so the three main senses are those with which
we are best to be concerned.

8.4.1 Vision

Changes in posture and colour are the main ways that animals communicate through the visual channel. Vision is the most important sense used
by predators that hunt by day, for it is the best for locating, and homing
in on, prey. This means that it is not the ideal sense for other animals to
use when advertising for mates. Visual signals also do not travel round
corners, so are not good for communicating in a dense environment such
as a forest, and the distance over which they can be transmitted is limited
by the size of the signaller: small animals cannot be seen a long way off.
For all these reasons, the main use of visual signals is in private, short-range
communication such as that between mates or between rivals disputing
at a territorial boundary (as in Figure 8.7). When not in use they tend to
be hidden so that the animal does not attract predators: most of the bright
colours on birds are on their undersides, butterflies fold their wings so that
their most colourful surfaces are hidden, and many fish have the capacity
to change their colour so that they are less conspicuous. The small pale
grey fish which is now hiding in the weeds from a passing pike is hardly
recognisable as the jet-black ten-spined stickleback which was chasing off
an intruder a few minutes ago. It is interesting that the most brightly
coloured animals are found amongst the insects, fish and birds. These,
through their capacity to fly or swim, can move in three dimensions and
so have extra scope for escaping from predators.

8.4.2 Audition

Sound is not as private a channel of communication as is sight, for noises
travel out from a source in all directions and cannot easily be limited, so
that they are only transmitted to one individual. Sounds can also pass

round corners and they move rapidly through the environment, though admittedly less so than visual signals travelling at the speed of light. Using sound is thus a good way of advertising, provided the animal producing the signal is large enough to generate sufficient noise (Figure 8.10*a*).

In addition, sound communication has the merit that a great deal of information can be transmitted very rapidly. The pattern of frequency in time which codes this information can be changed with great speed so that one signal can follow quickly on the heels of another to a much greater extent than would be possible with visual signals. Sound communication is therefore ideal not only for advertising over long range but also for other signalling, such as in our language, where a great deal of information is required to be transmitted very rapidly.

Communication by sound provides some nice examples where the exact form of the signal appears to be related to its function. Producing an alarm call in a form the predator cannot easily localise is a case we considered earlier. By contrast with this, advertising animals often produce broad frequency band sounds which are easy to locate and towards which others

Figure 8.10. Visual signals are easily localised and tend not to travel far, making them ideal for short-range courtship and aggressive signals, such as those of the fighting fish shown in Figure 8.7. Sounds travel rapidly over long distances so are often used by animals advertising for mates, as with the mole cricket illustrated in (*a*). In the species shown, the animal constructs a horn-shaped burrow which enhances sound transmission (redrawn from H. C. Bennet-Clark (1971), *Nature* **234**, 255–9). Smells have the advantage of persistence, so many mammals use them to mark territorial boundaries where they continue to act after the signaller has moved on. In (*b*) a blackbuck is shown marking a tree stump with secretions from a facial scent gland.

(*a*)

(*b*)

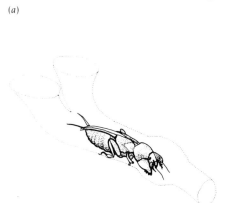

can orientate without difficulty. However, some sounds travel better than others through different environments, and the habitat in which a species lives may also influence the form of the signal it is best for it to produce. In woods, for instance, sounds echo off trees so that the pattern in time of a series of sounds becomes distorted. By contrast, pure tones can pass relatively uninfluenced through such an environment. As one might expect from this, bird species living in the tropical forests of central America use tones in their songs more than do the open-country species living nearby.

8.4.3 Olfaction

At first sight the production of olfactory signals, or pheromones as they are usually called, might seem to have very few advantages. Smells diffuse only slowly through the environment, their speed and direction of travel being highly wind dependent, and they can carry very little information, for after one smell is released time must elapse for it to disperse before another signal can be employed. However, there are situations where pheromones are ideal. A small animal such as a moth could not be seen or heard from more than 100 metres or so away, no matter how brightly coloured or noisy it was. Yet the male of some moth species can detect the pheromone produced by the female several kilometres away. The chemical involved is a small molecule so that it diffuses rapidly, yet large enough that its structure can be species-specific; the male need only sense a few molecules to start moving upwind to where the female waits for him.

Another merit of pheromones lies in their very persistence: they can continue to signal even when the animal is not there. Many mamals have different scent glands which they use for various purposes. Territorial species often mark the boundaries of their range as a signal to others that the area is occupied (Figure 8.10b). They may do this with special glands, or with scents in their urine or faeces, and a wealth of information can be contained in such signals. The potential intruder may glean, for example, not only where the territorial boundary is, but which individual is occupying it, what reproductive state he is in and even, by sensing how fresh the mark is, how long ago it was that the owner of the territory last went by.

Thus, though olfaction might seem to us the poor sister of the senses as far as communication is concerned, its very persistence and slowness of dissipation make it adapted for uses of its own for which the other senses would be of little use.

8.5 Animal communication as language

Most animal communication is very different from human language, as the messages it transmits are not at all precise and word-like. They are much more akin to the signals we communicate to each other with our faces: the yawns, smiles, laughs and frowns which indicate to other people what our feelings are. Language is different from this in many different ways. But the difference is not so great as might at first sight appear, for some animals are capable of transmitting detailed and accurate information just as we are. It is interesting that one of the best examples is a humble insect rather than an animal more closely related to ourselves. The dance of the honey-bees, first described by Karl von Frisch, and for long questioned because it seemed so extraordinary, is certainly one of the most remarkable forms of communication to have been discovered.

On a sunny summer day, worker bees fly out from their hive in search of the pollen and nectar on which the colony feeds. If a worker finds a good food source within about 100 metres or so of the hive she will return

Figure 8.11. The waggle dance of the honey-bee gives other workers information about the location of food. The nearer the source the more rapidly the dance is performed. The dance is in the form of a figure of eight and the orientation of the central part of this in relation to the vertical within the hive is the same as the angle between the direction of the sun and that of the food outside. The forager shown dancing here is doing so directly up the comb, thus indicating a food source in the same direction as the sun. (After K. von Frisch (1966), *The Dancing Bees*, 2nd edn, Methuen, London.)

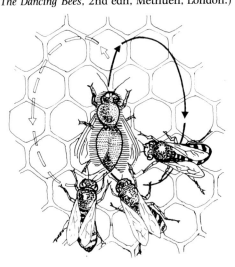

and perform a *round dance* on the vertical surface of the honey-comb. Her movements attract other workers and they crowd round her sensing the smell of the flower on which she has been feeding. They will then fly out and search for this smell in the neighbourhood of the hive, thus homing in on the food supply that she has found.

The round dance does little more than stimulate other bees to take an interest in the dancer, but the information in the *waggle dance*, which workers returning from more distant sites perform, is much more detailed (Figure 8.11). In this dance the bee performs a figure of eight, looping over the surface of the comb, first in one direction then in the other, and waggling her abdomen as she goes. What von Frisch discovered was that this pattern informs other bees of both the direction and distance of the source of food that the forager has discovered. Direction is indicated by the orientation of the dance upon the surface of the comb. If the food is directly towards the sun, the bee will make the middle run of its figure of eight vertically up the hive, and if in the opposite direction it will go straight down. In other words the angle between the sun and the source of food as the bee emerges from the hive is the same as that between vertical and the direction in which it dances when back inside the hive. A remarkable feature of this system is that workers need neither see the dancer nor the sun to be recruited. It is dark inside the hive so they can only sense the direction the dancer takes by touch. Outside the hive, the sun may be hidden behind cloud but the bees will still take up the correct direction as long as a small patch of blue sky is visible. This is because they are sensitive to the pattern of polarised light in the sky and can tell where the sun is from this.

So much for direction. It is the waggles that the forager makes as it dances that indicate distance. A worker that has had to fly a long way dances rather lethargically, whereas one that has found food close to dances with vigour and so produces more waggles per unit time. Each waggle is accompanied by a burst of sound and it is probably the rate of these sound pulses that is used by other bees to assess distance.

The bee dance has several notable attributes, two of which are features one might otherwise think of as unique to human language. One is that it is *symbolic*: information about distance and direction are encoded in features of the dance in a stylised way. The other point is that the bees are communicating, as we often do, about events which are distant in time and space from where the communication is taking place. In other words, in the darkness of the hive they are telling each other where the food is even though this may be some kilometres away.

Figure 8.12. A chimpanzee using American Sign Language (ASL). The chimp shown here is Moja, one of Washoe's successors. She is being questioned about a birch tree and has put her arms into the correct configuration to signal the word 'tree' (photograph by courtesy of B. T. and R. A. Gardner.)

The communication of bees is remarkable, but comparing it with human language is a somewhat sterile pursuit because the two systems are obviously vastly different in complexity. An even more contentious subject has been the comparison between our language and that which people have succeeded in teaching to chimpanzees and gorillas. Initial efforts to 'teach chimps to speak' were to no avail, but probably partly because they lack the vocal apparatus necessary to produce the sounds we can master. More success has been achieved with methods which depend on the skills they have with their hands. In the best known case, Allen and Beatrix Gardner taught a young female chimp called Washoe nearly 200 of the gestures used in ASL, the sign language used by the deaf in America (Figure 8.12). Washoe used them in appropriate contexts, strung them together in novel ways, could grasp concepts such as 'open' (applied to both a door and a matchbox) and 'dog' (applied to things as different as seeing a chihuahua and hearing a bark).

Whether Washoe's feats amount to language in the human sense has caused a great deal of heated argument which has shed next to no light on the issues. It will always be possible to point to ways in which such animals fail to match up to human achievements for the simple reason that they are not human. But perhaps two things are worth pointing out. First, whether one wants to label them as language or not, the communicative skills that chimps such as Washoe have achieved are obviously remarkable. Secondly, it is very unlikely that chimpanzees possess such capacities and yet leave them lying latent. In the wild they are sociable animals, living in rather loose groups. The sounds, gestures and facial expressions that they make may not appear complicated to us, but there may be a hidden wealth of meaning in them which will only be unravelled when we decide to invest as much in learning to understand their language as we have already done in trying to teach them to speak ours.

9

Social organisation

Strictly speaking, it is questionable whether one should refer to animal societies as 'organised' in any way. As earlier chapters have emphasised, natural selection acts on individuals, favouring those that are well adapted and whose genes thus spread at the expense of those of others, but not selecting for one group rather than another. Societies have thus not evolved through the action of natural selection upon them, but they have emerged as a result of the way selection has affected the behaviour of their individual members. Nevertheless, different animal species occur in very different social structures and, though each may be striving selfishly to maximise his inclusive fitness, elaborate societies may result from the way in which such individuals interact with each other.

In this last chapter we will consider some of the features of these societies, how they differ from each other, how they come to have a semblance of being organised and what social life involves for animals that belong to such groups. But first we must consider why animals would want to live in groups in the first place.

9.1 Why live in groups?

Some animals, such as tigers and other solitary hunters, spend most of their time on their own, though even they must come together in pairs for reproduction. Nevertheless, their mode of food finding is one with which the presence of others would interfere and their life is thus essentially a lone one, wandering widely over a range on which they hunt. Sometimes such 'home ranges' overlap so that different individuals may use the same area and, if those of males habitually overlap with those of females, they may meet for mating without either leaving its patch. Sometimes, however, animals live as individuals, pairs or larger groups on 'exclusive territories' from which they repel other members of their species, at least those of the same sex as themselves. The boundaries between such areas can be fiercely contested and often end up being quite sharp (Figure 9.1). A nice example

165

Figure 9.1. An example of exclusive territories. The symbols on this map show the positions within a wood in Sussex where 11 different male chaffinches were heard singing during a few weeks in spring. There is little overlap between the birds and, in the few places where this does occur, it is probably caused by boundaries shifting during the course of the study. (Data by courtesy of D. J. Goodfellow.)

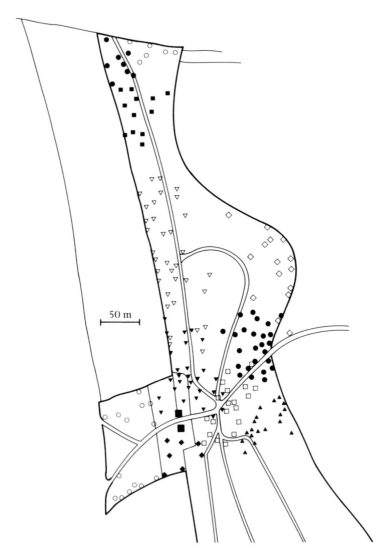

here, where the borderline can actually be seen, is in the mudskipper, a small fish which has the capacity to live out of water for periods when the tide is out. The males are highly territorial and build mud walls between the areas they occupy so that, by the time the tide returns, the mud is a mosaic of areas, each fenced off like a suburban garden. When the tide is out, the walls preserve some water within the territory and this helps the animal to respire, but the main function of the territory seems to be as a display area, for the males leap in the air and this attracts females to mate with them.

There are several possible reasons why animals might defend territories. The two most common ones are that it enables them to court and mate without interference from others and that it allows them to sequester a food supply which they can use systematically without others raiding it. The males of many small songbirds set up such territories, and these are usually large enough to provide enough food both for themselves and for their mates and offspring.

Defending a territory as a source of food is mainly a practicable proposition if the food supply is rather constant and evenly spread. If this is not so, then the area required to feed the animal, or a family if it is during the breeding season, may be so great that it is just impossible for one animal to patrol its boundaries and ensure that others are kept out. The distribution of food and the amount of it that an individual can defend can have a profound influence on social structure. For example, it has been argued that polygyny, a single male having more than one mate, may have evolved in territorial birds where some territories are very much richer in food than are others. As a result a female may breed more successfully by being the second one of a male on a rich territory rather than the sole partner of a male whose territory is poor. If food is patchy in time and space, this may also have implications for social structure. European badgers live in groups of 2–12, on ranges which also vary enormously in size. Their food is largely earthworms, and the abundance of these varies from place to place and also with time in the same place. The range size in a particular area is such as to ensure that there is always a productive patch somewhere within it, and the group size is related to how good a supply of worms there is overall within the range.

It is probably largely such considerations of food availability that determine whether animals maintain territories or live on rather looser home ranges. Whichever is the case, however, the numbers of individuals that occupy the territory or range may vary from one to a sizeable group, all moving around together and sharing the resources of the area, or all

Figure 9.2. Within a flock of feeding geese a few individuals have their heads up, scanning for danger, at any one time. Those that are foraging can thus do so safely as they will be warned should a predator approach. The larger the flock the less time each individual needs to spend with its head up (photograph © M. Owen, Wildfowl Trust, Slimbridge, Glos.).

foraging more or less independently of each other. Why should it benefit selfish individuals to live in such groups? Most of the reasons that have been suggested relate to ways in which grouping helps in the finding of food and in defence against predators.

9.1.1 Defence against predators

Animals in groups gain a number of advantages related to defence. While a group of animals is bigger than a single one, and so can be spotted from further off, if the predator only eats one prey at a time it will take longer to find a meal if the prey are clumped than if they are evenly spread out. Furthermore, there are more pairs of eyes in a group, so the predator is less likely to creep up undetected. To some extent this is offset by the fact that animals in groups are able to be less vigilant than solitary individuals.

Lone geese, for instance, raise their heads much more frequently while cropping the grass than they do when feeding in groups. Because there are more pairs of eyes they can afford to keep their heads down and feed for longer periods with the security that others will raise a warning if danger threatens (Figure 9.2).

Another reason for grouping was pointed out by William Hamilton in a paper called 'Geometry for the selfish herd'. He made the simple point that, if predators just take one prey at a time, the best prey strategy is to keep another individual between oneself and the predator (Figure 9.3). Prey animals that do this will tend to form groups simply because being in a group minimises the area of danger to each animal. Being in a group can also confuse a predator because it has difficulty fixating upon and chasing a single individual. Thus a pike put in a tank with a single stickleback will rapidly catch it but, if there is a shoal of sticklebacks, catching the first one takes much longer even though the sticklebacks do not combine to defend themselves actively in any way. Some animals do cooperate in defence, however, giving yet another advantage to group living. Musk oxen set upon by a pack of wolves will form a circle facing outwards so that the wolves are confronted by a solid wall of horns and cannot reach any of their vulnerable flanks.

A final example of grouping geared to predator defence is a very curious one and this is in the nesting behaviour of ostriches. Several female ostriches lay eggs in one nest, but only the first to lay incubates: she accumulates a much larger number of eggs in this way than she could lay herself. After the eggs hatch the female may have around 25 chicks following her as she sets off across the savannah, but later she may collect even more as a result of chasing other hens away from their chicks (Figure

Figure 9.3. To illustrate his idea of the selfish herd, Hamilton used the example of frogs sitting on the edge of a pond from which a snake sometimes pounced and took the one nearest to it. Each frog thus has a zone of danger which stretches half way to its nearest neighbour on either side. An individual can thus minimise its own threat by moving so that others are close to it on both sides, as the dark animal is shown to do here. If all of them do this, they will end up in a tight group. (After W. D. Hamilton (1971), *J. theor. Biol.* **31**, 295–311.)

Figure 9.4. An ostrich surrounded by many chicks. Some of these will
have hatched from eggs the female laid herself, some from eggs other
females laid in her nest, and some of them may have come from other
nests and joined the group later. The advantage to the adults in tending
these creches is thought to result from a dilution effect: if a predator
takes one chick, the more there are in the group the less likely is that
chick to be one of the adult's own (photograph © F. W. Lane.)

9.4). Brian Bertram, who studied this extraordinary system, even recorded
one group in which two adults were accompanied by 105 chicks.

What advantage can there be to a female ostrich in caring for the eggs
and chicks of others? The answer seems to be that chicks survive better
in larger groups so that her own chicks benefit by a 'dilution effect'.
Predators such as golden jackals will eat ostrich eggs and chicks, but the
more of these that a female has with her the less likely is the one that is
taken to be one of her own offspring. This cannot be the whole story,
however, otherwise a female would gain whether she cared for her young
herself or let another female do so. At the stage of incubation it seems that
there *are* advantages to the hen that sits, for she can recognise her own
eggs and she gives preference to them so that they remain in the centre
of the nest and are more likely to hatch. She is not therefore being a
generous nanny helping out others, but is running the crèche for

Figure 9.5. Grouping as an aid to feeding. By hunting in groups small carnivores, like the wild dogs shown here, are able to tackle prey much larger than a single animal could subdue (photograph by P. A. Jewell.)

her own advantage. It remains to be seen whether females with chicks also gain from giving care themselves rather than leaving them to be looked after by others.

9.1.2 Finding food

A further range of benefits to group living is related to the exploitation of food supplies. If food is clumped, as is very often the case, animals that feed on it tend to become so too. Seeing another individual feeding may be a good indication of where food is abundant, so groups may form for this reason alone without there being any advantage in being a member of a group. Indeed the animals may interfere with each other's feeding efforts but, if they are conspicuous, there is no way in which they can stop newcomers arriving to share in the spoils.

Not all feeding groups, however, are passive clumps. While many predators, such as frogs, snakes and domestic cats, are solitary hunters, using stealth to catch small prey, some others, like lions, spotted hyaenas and wild dogs, cooperate in the hunt (Figure 9.5). This may increase the chances of success, because individuals can take turns in chasing and because they can spread out and so limit escape routes. It also enables them to tackle far larger prey than could an individual on its own: a pack of hyaenas can pull down a zebra or a wildebeest, or some of them can distract a rhino mother while others attack her calf.

A final advantage of grouping may also be important in animals that are not carnivorous. This is that groups may be able to make use of resources in a more systematic way than can isolated individuals. The bumble-bee flying from flower to flower may well be following close behind another one and so getting less reward than it would otherwise do. If individuals form flocks and move round their range together, as is often true of small birds outside the breeding season, they can be sure that they are all arriving in a fresh area where the food has had a chance to replenish itself since last they came that way.

Many monkey groups occupy quite large territories and move over them in a single party searching for the fruit or buds on which they live. They too may gain from the systematic foraging that group living allows, but their groups are more integrated and the relationships of individuals within them more varied than in a flock of birds or a herd of deer. The advantages of food finding or predator avoidance may explain why they, like many other animals, come to live in groups, but do little to account for the richness of the interactions between the group members. It is to various facets of the behaviour of individuals within groups towards one another that we shall now turn.

9.2 Kinship

Not surprisingly, animals that live in social groups tend to be related to one another. Young ones are born into the group to which their mother and father belong and may often stay in it to become parents themselves. However, if this was always so, animals in the group would become more and more inbred. Too much inbreeding leads to young which are less viable, so it is discouraged by natural selection. What usually happens, therefore, is that, as maturity approaches, young animals of one sex or the other move away from the group in which they were born and so do not mate with close relatives. In birds it is usually the females that disperse to different areas, whereas in mammals the males are more likely to leave, although there are quite a few exceptions to this rule, including species in which both sexes move away.

Which sex disperses may have a profound effect on various aspects of social behaviour. Male white-crowned sparrows sing songs which are almost identical with those of their neighbours and quite different from those of birds a few kilometres away (Figure 9.6). Song learning usually only occurs in the first three months of life and so is complete long before they set up their territories and start to sing. But they breed only a short distance from where they hatched and thus the song dialects are maintained. Although female birds do not usually sing, their choice of mate is probably influenced by the songs of prospective partners. In white-crowned sparrows, for example, there is evidence that females may prefer

Figure 9.6. The songs of each of six male white-crowned sparrows from three different areas around San Francisco. There are marked dialects, most obviously in the structure of the second part of the song, which is closely similar within an area but different between them. (After P. Marler & M. Tamura (1964), *Science* **146**, 1483–6.)

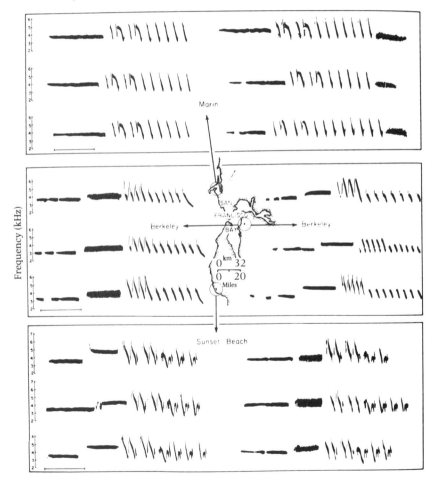

males from the same dialect area as themselves, though there is some controversy about this.

By contrast with birds, groups of many mammalian species are matri-archal, females staying with their mothers, while males leave to find other groups which they can join. Lions are a case in point here. As was mentioned in Chapter 7, males leave their natal group and go in search of another that they can take over. Females, however, stay in the group in which they were born so that the lionesses in a pride are usually closely related to one another. They often give birth at around the same time and it has been found that they will suckle each other's cubs, something one would only expect if they had an interest in them through being related. Furthermore, again in line with what kinship would predict, they suckle their sisters' offspring more than those of their cousins.

Alarm calling is another behaviour where kinship is important. Making a noise when a predator is around is obviously a risky thing to do, and one would expect to find some advantage to offset this. Many birds only call when they have young in the nest, so saving these from being discovered seems to be the benefit. Using a stuffed badger as a 'standard predator', John Hoogland has shown that black-tailed prairie dogs in America produce more alarm calls when they have close relatives in their group than when they do not. Males called a lot in their natal group but ceased to do so after they moved elsewhere, only to start calling again once their own young were born in the new group that they had joined. All these findings fit in well with the idea that animals are more likely to call when relatives may benefit.

As far as the relatedness of their members is concerned, the most complex groupings are amongst the social insects. In the honey-bee, only the queen lays eggs and all other females in her hive are sterile workers. The eggs the queen lays are of two sorts: unfertilised ones, which will develop into fertile males (drones), and fertilised ones, which usually develop into infertile females (workers) but, if nourished only on royal jelly, will form a new queen. This happens when the old queen dies or when the colony has grown to a point where it must split. A peculiarity of this breeding system is that drones have half the number of genes that queens or workers do and, when they fertilise an egg, *all* their genes pass to their offspring instead of just half of them. As a result of this system, which is known as haplodiploidy, workers share $\frac{3}{4}$ of their genes with their sisters rather than $\frac{1}{2}$, as most animals do. It is thought to be for this reason that it benefits them to raise sisters which will become queens rather than having daughters of their own. Through a quirk of their reproductive

Figure 9.7. A colony of naked mole rats packed tightly together in its burrow. The social system of this species is more like that of a social insect, with castes and adults that do not breed, than it is those of other mammals (photograph by J. U. M. Jarvis.)

system, their sisters are more closely related to them than their daughters would be and are thus a better investment in the future of their genes.

The social system of bees is thus extraordinarily intricate, and this is probably because these animals are especially closely related to others in the hive to which they belong. However, having this unusual mode of inheritance is not a prerequisite for a social system of this sort. Termites are not haplodiploid yet they have a colony structure which is just as complex. More remarkable still is the naked mole rat, for this is a mammal which has colonies much more like those of a social insect (Figure 9.7). These animals live communally in groups of around 40 individuals which share a burrow. For most of the time they huddle together in one underground chamber, and it is here that the only female to breed has her young. She is large, and males approaching her size may mate with her, but most other large animals in the group neither breed nor forage: their main role seems to be to keep the colony warm. Smaller individuals are like the workers of social insects. They nest build and move around through the burrow system foraging for roots and tubers which they bring back for all the animals to feed upon. If there is a disturbance, all the members of the colony join in the task of carrying the young off to safety. As well as the food that the workers bring back being shared, the young obtain nourishment from adults by eating their faeces. Individuals also spread their urine round the colony and groom with it. These habits, unsavoury as they may seem, probably serve to pass information around the group for, if the breeding female dies or is removed, another one comes into reproductive condition very soon. This is just like the social insects. In these, queen substance is spread throughout the colony and stops the workers from rearing other queens. But, if the queen dies, this substance is no longer there, and the workers quickly start to rear a replacement.

Mole rats are not very mobile and their burrow systems are isolated from each other. It is likely, therefore, that most of the animals within a colony are relatives and that their close social relationship stems from the common interest that this gives them in the offspring of the breeding female. Unfortunately we do not know just how closely related they are to each other, and this is very often the case in studies of social behaviour. It is easy to speculate that kinship is involved when we observe animals being altruistic to one another, but only in a few cases, such as that of the social insects, do we know in any detail exactly what the relationships are between the individuals involved.

9.3 Cooperation

Not all cooperation involves kinship. If two jackals help each other to capture a Thomson's gazelle that neither on its own could catch up with, then both of them show an immediate gain and the cooperation is of obvious advantage whether they are related or not. Rather more difficult to explain are cases where unrelated animals are generous without getting anything in return. A likely mechanism here, though one of which there are as yet very few examples, is referred to as 'reciprocal altruism'. This is where one animal helps another member of its group on the expectation that the favour will be repaid at some later date. It is just like cooperation, except that it involves a time lag. It is obviously a very important phenomenon amongst humans, for many of our relationships are based upon kindnesses which, while they may run in one direction in the short term, balance out in the long run.

For reciprocal altruism to evolve there are two important prerequisities. First, animals must stay together for long enough for it to be likely that an opportunity for repayment will arise. Secondly, animals must be able to recognise each other as individuals. If they cannot do so then 'cheaters', animals that receive but refuse to give in return, will remain undetected.

Figure 9.8. A vampire bat. In this species an individual will sometimes share the blood on which it has fed with another group member. This benefits both of them as the donor will usually be repaid by the recipient on a later occasion (photograph by U. Schmidt.)

As these get all the benefits without bearing any of the costs they will be at an advantage unless they can be excluded from the system. Given these points, it is not surprising that few examples of reciprocal altruism have been described in animal groups. However, a clear case was found by Craig Packer amongst baboons. In troops of these animals, one male, the alpha individual, is usually dominant and he mates with the most receptive females. However, Packer found that pairs of young males would sometimes cooperate, one of them distracting the dominant animal while the other mated with his consort. This was obviously of benefit to the one that mated, but what was in it for the other? The answer lay in reciprocation. At some later date the two would cooperate again, but on this occasion it would be the second that mated while the first did the distracting. In the long run, therefore, both would benefit.

The help which animals living in groups often show towards one another may be advantageous in several different ways. If two animals cooperate in some task such as prey capture both may show an immediate gain. But, where one animal assists another without any immediate profit to itself, it is likely either that the individuals are relatives, so that assisting benefits the giver's genes, or that reciprocal altruism is involved and the favour will be repaid at a later date. Jerry Wilkinson has found that both these factors are important in the food-sharing of vampire bats (Figure 9.8), a species that would not come to mind as an obvious example of an altruist! These animals live in groups and, if an individual fails to find a blood meal one night, another group member will regurgitate blood for it and so save it from starving. Usually the two animals in this arrangement are kin but, where they are not, they are most likely to be the ones in which the recipient gave blood to the donor on a previous occasion. Thus, to explain the food-sharing of vampire bats, it is necessary to invoke both kinship and reciprocal altruism.

9.4 Dominance

It is a rare group of animals in which all is peaceful and harmonious, each assisting the others and food being shared equally around. More often fights and squabbles occur and some animals do rather better than others, getting more of the food and doing more of the mating than their companions. Where one animal habitually beats another in fights, the first is said to be dominant and the second subordinate. In many animal groups, one individual, usually a large and strong male, dominates all others. This alpha male has priority of access to any choice food that the group comes across and to females within it that are receptive.

An interesting feature of dominance is how clear cut it often is, the alpha individual always defeating all others, the beta all but the alpha, and so on in a straight hierarchy down to the most lowly individual who wins no fights at all (Figure 9.9). This has been found even in quite large groups of chickens, though the occasional triangle may develop in their peck-order, so that A beats B, B beats C and C beats A. If a new group of animals is set up, they tend to squabble at first, but after a little while they settle down, each of them knowing his place, chasing off those beneath him but not prepared to challenge those above. Once established, the hierarchy thus persists more through deference on the part of subordinates than through actual fighting. Dominant individuals are often older ones, or ones that are bigger and stronger for some other reason, so that if there was a fight they would be likely to win. After some initial sparring, however, the others learn that it is not worth initiating such a fight. In small groups, once an animal is defeated, it may recognise the particular individual that beat it when next they meet and know from past experience that he is to be respected. But, where groups are larger or animals do not have the capacity to recognise one another, they may still form hierarchies simply by assessing each other's fighting ability, smaller and weaker individuals tending to give way to those that are more intimidating. Indeed the sparring and threatening performed by animals when displaying aggressively may be very much a matter of sizing each other up and assessing their relative fighting abilities. Some fish have 'pushes of war' and these are extremely sensitive to size differences: the bigger one wins even if the difference in size is very slight. Similar fights occur in deer, the stags interlocking their antlers and pushing for all they are worth (Figure 9.10). In the spider *Agelenopsis aperta*, individuals almost literally weigh each other. In environments where web-sites are scarce, an intruder may try to take over a site from its owner. If they are equally matched, the owner usually wins the encounter, but a large spider can often oust a smaller one from its web. Susan Riechert, who has studied these spiders in New Mexico, found that intruders with lead weights attached to them were more likely to be successful: apparently owners are deceived into giving up because

Figure 9.9. Dominance hierarchy between those seven workers in a colony of *Leptothorax allardycei* ants which had the most encounters with each other during an observation session of 18 hours. The figures show the number of times that the animal on the left defeated that at the top. During an encounter one animal pummels another, which crouches and freezes until the victor moves away. In these ants the hierarchy is a simple linear one with each ant always being dominated by those above it and dominating those below. (Data from B. J. Cole (1981), *Science* **212**, 83–4.)

Dominant				Subordinate			
	oe	he	po	sa	lo	mi	lp
oe	—	43	19	18	1	2	2
he		—	45	20	3	1	5
po			—	3	1	5	3
sa				—	2	3	2
lo					—	1	0
mi						—	2
lp							—

Figure 9.10. The sparring of red deer stags. By such pushing contests animals may be able to assess their own strength relative to that of others with great accuracy (photograph by G. Lincoln.)

they assess the danger by the amount of tension the rival produces in their web.

The advantages of being a dominant male need hardly be spelt out. A red deer stag which can fight off other males may hold a harem of many hinds and, in a few short weeks, father many offspring. But why do individuals lower down acquiesce rather than fight their way up? This is probably largely a matter of 'living to fight another day'. As they are often smaller or younger, it is to the advantage of these subordinate animals to bide their time, putting on weight and gaining in experience until they are ready to challenge the individual at the top. No one lives for ever and, in the case of the red deer, the stag with the harem has a particularly short expectation of life. Not only does he exhaust himself by mating repeatedly, but he must endlessly round up his hinds and ensure that any other males are driven off. Even if he is not actually injured in fights, he tends to lose condition rapidly during the rutting season and his chances of survival thereafter are much reduced.

In most species males are bigger, stronger and more aggressive than females and, if both sexes form a single hierarchy, as happens in a group of weaver birds, the females tend to be dominated by the males. But it is not always the case that hierarchies are straightforward listings of animals in order of size or strength. Sometimes one animal may beat another on some occasions and be beaten by it on others. This may depend on exactly what they are fighting over, but the circumstances leading to such 'back pecks' are not yet worked out. In other cases the genetic relationships between animals in a group may affect their status. Rhesus monkeys live in matriarchal groups, young females staying with their mothers to breed while their brothers disperse elsewhere. Several generations may occur in the same group and, in this case, the older females are dominant to their daughters and grand-daughters. Within a generation, however, dominance is in inverse order of age, a female's youngest daughter coming to dominate her older sisters as soon as she is mature. She is able to do this partly because her mother manages the situation and shows a certain amount of favouritism to her last-born.

Dominance hierarchies are a feature of animal societies that gives them the impression of being highly organised for, at least in cases where they are clear cut, each individual knows its place, defers to those above and ousts those below. Open aggression is rare once positions are settled and a threat display, or perhaps just a slight movement of intent, may be sufficient for the superior to assert its position. As with other aspects of social behaviour, however, the organisation is more apparent than real. The hierarchy exists because each individual has discovered that he is able to dominate those below him but not those above.

9.5 Synchrony

If animals are to stay in the groups to which they belong they must do roughly the same thing at the same time as each other. It is safe to groom or to sit quietly when others are sleeping, but the animal that falls asleep when the group is foraging or starts to explore while the rest are asleep will quickly find itself alone.

Synchrony of behaviour is achieved in a number of ways. One of these is that, quite simply, animals tend to do certain things at a particular time of day even if entirely on their own (Figure 9.11). Most species that are active by day show a burst of moving around and looking for food in the hours immediately after dawn, they will then rest and groom during the mid-day period, and perhaps show a brief resurgence of activity just before dusk. Nocturnal species tend to show the opposite pattern: a peak of activity just after the sun goes down and a more minor one just before dawn.

Patterns like these, on their own, will lead to different animals showing broadly similar behaviour at equivalent times. But the phenomenon of social facilitation sharpens up such timing. If a hungry rat is given food in the presence of a companion that has eaten its fill, the second animal is likely to start eating again, albeit in a rather desultory way. If one bird is given water, others able to see it bathing will start to wash themselves even though their water has been unchanged for days. And we all know how infectious yawning is! Social facilitation is a potent force leading animals to show the same behaviour as their companions.

It is perhaps curious that leadership is not usually involved in the synchrony of behaviour that animals show. Large and dominant individuals may determine when a group moves and when it rests more than others, but much of the patterning of group behaviour depends more on consensus than on despotism. The goose which takes the lead as they fly off in V-shaped formation is a different bird from one flight to the next. Even in primates, where the benefits of age and experience might suggest that it would

Figure 9.11. The behaviour of animals can change dramatically from one time of day to another. In species which are active by day, various activities tend to peak early in the morning, with a lesser peak as dusk approaches and a period of quiescence around mid-day. This is illustrated here for the flying activity of a tsetse fly. In this case the activity also mounts from day to day as time passes since the animal's last meal. All animals in a species tend to show similar changes, so the behaviour of different individuals will be synchronised without them necessarily being influenced by one another. (After J. Brady (1981), in *Handbook of Behavioral Neurobiology*, vol. 4, ed. J. Aschoff, pp. 125–44, Plenum, New York).

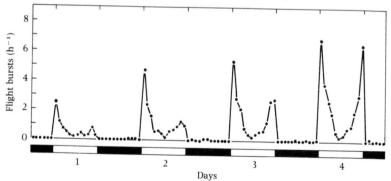

Figure 9.12. Many birds, such as the starlings shown here, can roost in huge numbers. These gatherings may act as 'information centres', so that individuals that failed to feed well on one day can find a better supply on the next by following others as they fly out in the morning (photograph by C. J. Feare).

pay to follow some individuals more than others, patterns of movement are far from well ordered. As a troop of yellow baboons moves along, the order of individuals in it is essentially random, though there is some tendency for both adult males and juveniles to be in the front third. In hamadryas baboons, where groups leaving the cliffs on which they sleep can be very large, Hans Kummer has described how different individuals may start off in different directions and then wait to see whether others follow. While the group may not have a single leader, some individuals may exert more influence than others by refusing to follow where they are going.

9.6 Culture

An advantage to animals that live in groups is that they can glean information from each other. Some bird species, such as starlings, gather in huge roosts each night (Figure 9.12). Being in a large group probably helps to warn them of the approach of nocturnal predators such as cats and foxes. But another benefit that they may gain is in information about where the best sources of food are to be found. It has been suggested that animals that have fed badly one day, by watching others departing from the roost next morning, may join a group which is going back to a place which the others had found profitable. Roosts may therefore act as information centres for those individuals that have been less successful at finding food than their companions and, as a result, the next day they may do better. In more permanent groups than those of roosting birds, individuals may benefit in an equivalent way. In times of drought, the most aged member of a monkey troop may have information about water holes that do not dry out but that others would be too young to remember.

It is not just information about possible food or water sources that passes from one animal to another within a social group. Young animals, like birds learning their songs or kittens mastering the skills of prey capture, may benefit from the example set by adults. Within a social group, such cultural influences are of enormous potential importance for, once a particular skill or piece of information is discovered, there is no reason why it should not spread through the group and persist long after its discoverer is dead.

The way in which blue tits and great tits in the British Isles learnt to open milk bottles (Figure 9.13) is a good case in point here. These birds are well equipped for the task, as they are attracted to bright objects and they gain access to sources of food such as nuts by hammering at them with their beaks. All it needed was for one of them to peck at a bright and shiny milk bottle top and, hey presto!, there was a splendid drink of cream. But the remarkable feature of this innovation was that it spread rapidly through the population so that, within a few years, no milk bottle in Britain was safe. Mapping the spread of the habit showed that it did not arise by chance but moved out from certain focal points. It is clear that some birds developed the habit for themselves and others amongst their social companions followed suit, either through copying or discovering bottles others had opened, so that it soon became widespread. Humans have now developed a counter adaptation, by putting empty yoghourt pots on top of their full milk bottles, and this is spreading through their population in an equivalent way. Perhaps the tits will develop the skill of flicking these off!

Another notable example of cultural transmission has been described in the feeding habits of Japanese monkeys. In one troop of these animals, a particularly innovative young female called Imo devised a number of new habits. She discovered that sweet potatoes tasted better if they were washed before eating. She also found a good way of dealing with grain which had been spread on the shore for her troop to eat. Other troop members laboriously picked out the grains one by one from the sand on which they lay. Imo, however, found that it was best to take a pile of sand and grain down to the lake and drop it in. The sand sank and the grain floated so that it could be scooped from the surface. In this way she obtained a whole mouthful at a time rather than having to pick it up grain by grain. Imo's

Figure 9.13. A classic case of cultural transmission: the opening of milk bottles by titmice. Here a great tit is shown removing the shiny top from a bottle before drinking the cream (photograph © R. Thompson).

skills did not die with her. Other animals in her troop copied what she did until, eventually, all but the very young and the very old behaved in the same way. The young had not yet had a chance to learn and the old were presumably too set in their ways. Nevertheless, these habits became widespread within the troop to which she belonged whilst being unknown amongst monkeys elsewhere.

The notion of cultural transmission as an important way in which animals that live in groups may obtain information is a good point at which to end. If there is any feature of ourselves that sets us apart from the rest of the animal kingdom it is the enormous importance that such culture plays in our lives. Reading books about animal behaviour and learning to wash sweet potatoes are obviously very different but, deep down, they have quite a bit in common.

SELECTED READING

The list given here is restricted to recent books which should be readily available through libraries. As their titles suggest, some of them are wide ranging, covering much of the same ground as this book but at a more advanced level, while others are relevant largely to one or a few of the topics dealt with here. The aim is to enable the interested reader to take the subject further rather than to provide an exhaustive list of sources.

Alcock, J. (1984). *Animal Behavior: An Evolutionary Approach.* 3rd Edition. Sinauer, Sunderland, Massachusetts.

Barnard, C. J. (1983). *Animal Behaviour: Ecology and Evolution.* Croom Helm, London.

Bonner, J. T. (1980). *The Evolution of Culture in Animals.* Princeton University Press, Princeton.

Brown, J. L. (1975). *The Evolution of Behavior.* Norton, New York.

Chalmers, N. R. (1979). *Social Behaviour in Primates.* Edward Arnold, London.

Daly, M. & Wilson, M. (1983). *Sex, Evolution & Behavior,* 2nd Edition. Willard Grant, Boston.

Dawkins, R. (1977). *The Selfish Gene.* Oxford University Press, Oxford.

Edmunds, M. (1974). *Defence in Animals.* Longman, London.

Gould, J. L. (1982). *Ethology.* Norton, New York.

Halliday, T. R. & Slater, P. J. B. (eds.) (1983). *Animal Behaviour.* 3 volumes. Blackwell Scientific Publications, Oxford.

Hinde, R. A. (1970). *Animal Behaviour.* McGraw Hill, New York.

Hinde, R. A. (ed.) (1972). *Non-verbal Communication.* Cambridge University Press, Cambridge.

Hinde, R. A. (1982). *Ethology.* Oxford University Press, Oxford.

Huntingford, F. A. (1984). *The Study of Animal Behaviour.* Chapman & Hall, London.

Krebs, J. R. & Davies, N. B. (1981). *An Introduction to Behavioural Ecology.* Blackwell Scientific Publications, Oxford.

Krebs, J. R. & Davies, N. B. (eds.) (1984). *Behaviour Ecology. An Evolutionary Approach,* 2nd Edition. Blackwell Scientific Publications, Oxford.

Manning, A. (1979). *An Introduction to Animal Behaviour,* 3rd edition. Edward Arnold, London.

Sebeok, T. A. (ed.) (1977). *How Animals Communicate*. Indiana University Press, Bloomington.

Smith, W. J. (1977). *The Behavior of Communicating*. Harvard University Press, Harvard.

Toates, F. M. (1980). *Animal Behaviour. A Systems Approach.* Wiley, Chichester.

Wickler, W. (1968). *Mimicry in Plants and Animals*. Weidenfeld & Nicholson, London.

Wilson, E. O. (1975). *Sociobiology. The New Synthesis*. Belknap Press, Harvard.

LIST OF SPECIES NAMES

babbler, Arabian, *Turdoides squamiceps*
baboon, hamadryas, *Papio hamadryas*
baboon, yellow *Papio cynocephalus*
badger, European, *Meles meles*
badger, American, *Taxidea taxus*
bat, fringe-lipped, *Trachops cirrhosus*
bat, vampire, *Desmodus rotundus*
bear, polar, *Ursus maritimus*
bee, honey, *Apis mellifera*
bee-eater, red-throated, *Merops bullocki*
blackbird, European, *Turdus merula*
blackbird, redwinged, *Agelaius phoeniceus*
blackbuck, *Antelope cervicapra*
bittern, *Botaurus stellaris*
bullfrog, *Rana catesbeiana*
budgerigar, *Melopsittacus undulatus*
bumblebee, *Bombus* sp.
canary, *Serinus canaria*
cat, domestic, *Felis domestica*
chaffinch, *Fringilla coelebs*
chimpanzee, *Pan troglodytes*
colobus monkey, black and white, *Colobus guereza*
colobus monkey, red, *Colobus badius*
cormorant, *Phalacrocorax carbo*
coyote, *Canis latrans*
cricket, mole, *Gryllotalpa vineae*
deer, red, *Cervus elaphus*
dog, domestic, *Canis familiaris*
dog, African wild, *Lycaon pictus*
dove, Barbary, *Streptopelia risoria*
duck, goldeneye, *Bucephala clangula*
duck, Mandarin, *Aix galericulata*
duck, pintail, *Anas acuta*
duck, wood, *Aix sponsa*
finch, zebra, *Taeniopygia guttata*
fish, Siamese fighting, *Betta splendens*
fish, electric, *Gymnarchus* sp.
fish, paradise, *Macropodus opercularis*
flatworm, *Dendrocoelum* sp.

fly, tsetse, *Glossina morsitans*
flycatcher, pied, *Muscicapa hypoleuca*
fox, red, *Vulpes vulpes*
frigatebird, *Fregata aquila*
frog, common, *Rana temporaria*
frog, North American cricket, *Acris crepitans*
frog, gray tree, *Hyla versicolor*
fruit fly, *Drosophila* sp.
gannet, *Sula bassana*
gazelle, Thomson's, *Gazella thomsoni*
gibbon, *Hylobates* sp.
giraffe, *Giraffa camelopardalis*
goldfish, *Carassius auratus*
goose, greylag, *Anser anser*
gorilla, *Gorilla gorilla*
grebe, great crested, *Podiceps cristatus*
guinea pig, *Cavia porcellus*
gull, black-headed, *Larus ridibundus*
gull, Galapagos swallow-tailed, *Larus furcatus*
gull, herring, *Larus argentatus*
guppy, *Poecilia reticulata*
hawk, zone-tailed, *Buteo albonotatus*
hawkmoth, eyed, *Smerinthus ocellata*
hen, domestic, *Gallus gallus*
heron, green, *Butorides virescens*
horse, *Equus caballus*
hyaena, spotted, *Hyaena hyaena*
jackal, golden, *Canis aureus*
jay, Florida scrub, *Aphelocoma coerulescens*
killifish, common, *Fundulus heteroclitus*
kittiwake, *Rissa tridactyla*
kob, Uganda, *Kobus kob*
leopard, *Panthera pardus*
lion, *Panthera leo*
mallard, *Anas platyrhynchos*
mantis, praying, *Parastagmatoptera unipunctata*
marmoset, common, *Callithrix jacchus*
mockingbird, *Mimus polyglottos*
mole rat, naked, *Heterocephalus glaber*
monkey, Japanese, *Macaca fuscata*

monkey, rhesus, *Macaca mulatta*
monkey, vervet, *Cercopithecus aethiops*
mouse (laboratory), *Mus musculus*
mudskipper, *Periophthalmus* sp.
mynah, Indian hill, *Gracula religiosa*
octopus, *Octopus vulgaris*
ostrich, *Struthio camelus*
owl, barn, *Tyto alba*
ox, musk, *Ovibos moschatus*
oystercatcher, *Haematopus ostralegus*
pelican, *Pelecanus occidentalis*
pigeon (rock dove), *Columba livia*
pike, *Esox lucius*
plover, *Charadrius* sp.
plover, ringed, *Charadrius hiaticula*
prairie dog, black-tailed, *Cynomys
  ludovicianus*
rabbit, *Oryctolagus cuniculus*
rat (laboratory), *Rattus norvegicus*
rattlesnake, *Crotalus adamanteus*
rhinoceros, *Ceratotherium simum*
robin, European, *Erithacus rubecula*
scorpionfly, *Hylobittacus apicalis*
silkmoth, *Hyalophora cecropia*

sparrow, house, *Passer domesticus*
sparrow, white-crowned, *Zonotrichia
  leucophrys*
sparrowhawk, *Accipiter nisus*
starling, *Sturnus vulgaris*
stickleback, ten-spined, *Pygosteus pungitius*
stickleback, three-spined, *Gasterosteus
  aculeatus*
sunfish, bluegill, *Lepomis macrochirus*
tit, blue, *Parus caeruleus*
tit, great, *Parus major*
tit, long-tailed, *Aegithalos caudatus*
toad, common, *Bufo bufo*
toad, South African clawed, *Xenopus laevis*
warbler, garden, *Sylvia borin*
warbler, marsh, *Acrocephalus palustris*
warbler, sedge, *Acrocephalus schoenobaenus*
water boatman, *Notonecta glauca*
weaverbird, red-billed, *Quelea quelea*
whale, killer, *Orcinus orca*
wildebeest, *Connochaete taurinus*
wolf, grey, *Canis lupus*
woodpecker, black, *Dryocopus martius*
zebra, plains, *Equus burchelli*

INDEX

action-specific energy, 52, 53
activity, 68, 103–5, 181, 182
after-discharge, 17, 18
aggression, 72, 136
  functions, 56, 57
  motivation, 54–9
  releaser, 59
alarm calls, 146, 147, 155–7, 174
altruism, 136–8
  kinship, 176
  reciprocal, 176–8
American Sign Language (ASL), 163
amphibian
  hearing, 46, 47
  vision, 41–6
ant (*Leptothorax allardycei*), 179
approach–avoidance conflict, 73, 79
audition, 158–60; *see also* hearing *and*
  communication

babbler, Arabian, 136, 137
baboon, 177, 183
baby, human, 97, 98
badger, American, 174
badger, European, 167
Baerends, Gerard, 39–41
Bastock, Margaret, 101–2
bat
  fringe-lipped, 145
  hearing, 35
  vampire, 177, 178
bathing, 114
bear, polar, 148
bee, 148; *see also* honey-bee *and*
  bumble-bee
bee-eater, red-throated, 136, 137
behaviour
  control, 25–8
  convergence, 106, 107
  development, 76–99
  evolution, 100–18
  function, 122–9
  genetic basis, 11
  orientation, 29–33

priorities, 65–71
sexual, 63, 68
social development, 95–8
social organisation, 165–86
species-typical, 21, 22
stereotyped, 20
synchrony, 181–3
systems, 68, 69
Bertram, Brian, 169–70
bird
  foraging, 133
  hearing, 34, 35
  song, 62, 63, 82
    development, 88–93
    function, 119–22, 124, 125
bittern, 145
blackbird, 146, 147
  European, 146
  redwinged, 141
blackbuck, 159
budgerigar, 122
bug detector, 43
bullfrog, 46, *see also* frog
bumble-bee
  foraging, 129–33, 142, 172
  social system, 12, 13

call notes, 155
canary, 93, 122
cat, 57
cat display, 117
causal factors, 68, 69, 72, 73
causation, 7–10
chaffinch, 82, 88–93, 166
chick, 95, 96
chicken, 178
chimpanzee, 163, 164
claustrophobic (fruit-fly mutant), 104
cliff-nesting, 125, 126, 127
Clutton-Brock, Tim, 127–9
cock, fighting, 72, 73
colouration, 147, 148
communication, 144–64, 183, 184
  analysis, 154–7

191

communication (*cont.*)
  costs, 145
  misleading, 147, 148, 149, 150, 151
conditioning, 77, 78
conflict, 73, 74
contact calls, 155, 157
cooperation, 176–8
copulation, 125
cormorant, 108
courtship, 23, 72
  feeding, 110
coyote, 114, 115
cricket, 48, 49
  mole, 159
crossed-extensor reflex, 25
crypsis, 144–50
Csányi, Vilmos, 94, 95
Cullen, Esther, 125, 126, 127
cultural transmission, 183–6

Darwin, Charles, 1
Davies, Nick, 152, 153
deception, 151
deer, 172
  red, 179, 180
deprivation experiment, 74, 83
description, 15
development, 7–10
disinhibition, 73
displacement activity, 67, 72–4, 114
display, 4, 5
  communal, 134, 135
  constancy, 116
  evolution, 107
  origin, 112–18
dithering, 71
dog, 81, 82
  conditioned reflex, 77
  olfaction, 35
  prey capture, 24
  reflexes, 16–18
  wild, 171, 172
dominance, 178–81
dove, Barbary (ring), 19, 69–71
drinking, 65, 66
  motivation, 54
  species-typical, 21, 22
drives, 66
Drosophila, see also fruit-fly
  courtship, 101, 102, 111–12
duck display, 107, 109
  goldeneye, 21, 79, 81
  mallard, 72, 73
  mandarin, 113, 114
  pintail, 107
  reproductive strategy, 138
  wood, 79–81

duckling, visual cliff behaviour, 79, 81
Dundreary attitude, 117

eating, *see* feeding
eclosion hormone, 27
egg-rolling, 23, 24
emancipation, 116
energy models, 54–8
ethogram, 3
ethology, definition, 1
evolution, 11–15, 142
evolutionarily stable strategy (ESS), 142
  mixed, 142, 151
Ewing, Arthur, 104–5
exhaustion, 54, 59–62
experimentation, 14, 15
external factors, 62–5
faeces, 160
fanning, 4, 5
fatigue, 17, 18
feedback, 24, 26, 58
feeding, 69–71
  motivation, 25, 54, 58, 63, 68
fighting, 135, 136; *see also* aggression
finch
  nest-building, 83
  zebra, 65, 66
fire-fly (*Photuris*), 149
fish
  aggression, 55, 56
  cichlid (*Haplochromis burtoni*), 56
  electric (*Gymnarchus niloticus*), 36, 37
  paradise, 95
  Siamese fighting, 151, 152
  territory, 167
fitness, inclusive, 138
fixed action pattern (FAP), 5, 20–5, 27, 28
flatworm (*Dendrocoelum* sp.), 31
fly
  empidid (*Hilara sartor*), 109, 110
  larva (*Sarcophaga barbata*), 19, 20
  tsetse, 182
flycatcher, pied, 139, 140
food
  finding, 171, 172
  species preferences, 93, 94
  territorial distribution, 167
foraging, 129–33
fowl, domestic, 95, 96, 178
fox, red, 147, 148, 154
frigate bird, 108
Freud, Sigmund, 96
Frisch, Karl von, 1, 2, 161, 162
frog, 169; *see also* bullfrog
  gray tree, 47
  North American cricket, 47

frog (*cont.*)
  Panamanian (*Physalaemus pustulosus*), 145
  senses, 41–7
fruit-fly, 101–5
function, 11–15
  concept of, 119–22

gait, species-typical, 21, 22
ganglion cells, 41–6
gannet, 108
Gardner, Allen and Beatrix, 163
gazelle, Thomson's, 176
genetics, behaviour, 100–6
Gerhardt, Carl, 47
gibbon, 138
giraffe, 21, 22
goldfish, 95
goose
  feeding, 168, 169
  flight, 182, 183
  greylag, 23, 24
gorilla, 163
grebe, great crested, 117
grooming, 63, 65, 66, 69, 73, 74, 114
Gryllus campestris, 48
guinea pig, 73
gull
  black-headed, 123, 124
  Galapagos swallowtailed, 128
  herring, 39–41
  see also kittiwake
guppy, 63, 64

habituation, 60, 61
Halliday, Tim, 152, 153
Hamilton, William, 137, 138, 169
haplodiploidy, 174
Harlow, Harry, 96–8
hawk, zone-tailed, 148
hawk moth, eyed, 147
head scratching, 22
head-toss display, 21, 116
hearing, 34, 46, 47, 48; *see also* audition
Hebb, Donald, 88
Heinrich, Bernd, 132, 133
Heinroth, Oscar, 1
heredity, 78, 87, 88·
heron, green, 113
hierarchy, social, 178–81
honey-bee
  breeding system, 174
  dance language, 161, 162
  social system, 12, 13
Hoogland, John, 174
horse, 21, 22
hoverfly, 148

Huxley, Julian, 116, 117
hyaena, spotted, 171, 172

inbreeding avoidance, 172–4
infanticide, 136
information centres, 183, 184
Innate Releasing Mechanism (IRM), 39
insect
  foraging, 132
  stick, 145
  vision, 35, 36
instinct, 66, 79–83
internal factors, 62–5
isolation, 55

jackal, golden, 170
jay, Florida scrub, 136, 137

Kaspar Hauser, *see* deprivation experiment
Kear, Janet, 79–81
Kessel, E. L., 109–10
killifish, 85
kin selection, 137
kinesis, 29, 31
kinship, 172–6
kitten, 84, 85
kittiwake, 125, 126, 127
knee-jerk reflex, 16, 17
kob, Uganda, 135
Krebs, John, 125
Kummer, Hans, 183

Lack, David, 58, 59
landmark orientation, 32, 33
language, 161–4
latency, reflex, 17, 18
leadership, 182–3
learning, 9, 76–9
  bird song, 88–93
  food preferences, 93, 94
  selection experiments, 105, 106
Lehrman, Danny, 83–8
leopard, 155, 156
lion, 136, 171, 172, 174
Locke, John, 76
Lorenz, Konrad, 1, 2, 20, 52, 53, 107, 108

magnetic field, 31, 32
Manning, Aubrey, 103–4
mantis, praying, 24, 25, 145, 149
marmoset, 138
mate
  attraction, 121, 122
  choice, 173, 174
  desertion, 139, 140
matriarchal social grouping, 174

Maynard Smith, John, 117
McFarland, David, 69–71
migration, 28–31
mimicry
  bird song, 92
  bluegill sunfish, 151
  hoverfly, 148
  insect, 145
mocking bird, 92
monkey
  colobus species, 127–9
  feeding, 172
  Japanese, 184, 185
  Rhesus, 97, 181
  troop information exchange, 184
  vervet, 155–7
Morris, Desmond, 116
motivation, 51–75
  fatigue, 18, 19
  models, 52–65
  theories, 10
motor patterning, 16–33
mouse, 55, 56, 62, 63, 102
movement
  intention, 113
  orientation of, 28–33
  peripheral control, 25–7
  protective, 115
mudskipper, 167
murder, 136
mutants, 101–3
mynah, Indian hill, 91, 92

natural selection, 11, 86
nature/nurture, 8
nest-building, 83
neurobiology, 10
Nobel Prize, 1, 2

observation, 13–14, 15
octopus, 35
oestrogen, 67, 68
olfaction, 159, 160
optic tectum, 43, 45, 46
optimality, 129–34
ostrich, 169–71
owl, barn, 34, 35
ox, musk, 169
oyster catcher, 22

Packer, Craig, 177
pain, 82
paternity uncertainty, 140, 141
Pavlov Ivan, 77, 78
pelican, 108, 109
pigeon, homing, 31, 32
pike, 169

plover, 22, 147,
  ringed, 148
polygyny, 167
posture, 158
prairie dog, 174
predation
  avoidance, 123, 124
  fire-fly, 149
predator
  camouflage, 148, 149
  defence, 168–71
  recognition, 93–5
prey capture, 24
  cooperation, 176, 177
  skills, 184
prey recognition, 93–5
progesterone, 9
psychohydraulic model, 52, 53
psychology
  Behaviorist, 76–9
  comparative, 7–10

rabbit, 24
rat
  feeding, 55
  food preference, 93, 94
  learning, 77, 87, 105, 106
  mating, 61
  naked mole, 175, 176
  social facilitation, 182
  urination, 83
rattlesnake, 35, 37
reflexes, 16–20, 77
refractory period, 61
releasers, 37–9, 66
reproductive strategies, 138–43
reproductive stimulation, 122
retina, 42
rhinoceros, 172
Riechert, Susan, 179, 180
ritualisation, 116, 117
robin, European, 31, 32, 59
roosting, 183, 184
round dance, 162

scorpion-fly, 110
scratch-reflex, 16–18
selection experiments, 11, 105, 106
sensitisation, 61
sensory systems, 34–50
shaking display, 117
Sherrington, Charles S., 16–17
sign stimulus, 37–9, 45, 46
signals, 155, 157–60
silkmoth, 27, 28
Skinner, B. F., 77–9
Skinner box, 69–71

Smith, John Maynard, 117
Smith, W. John, 154
snake, garter (*Thamnophis elegans*), 93, 94
social facilitation, 182
sociobiology, 134–8
song learning, 88–93, 184
sound
  imprinting, 95
  peripheral filter, 47
sparrow
  house, 22
  white-crowned, 172, 174
sparrowhawk, 146
spider (*Agelenopsis aperta*), 179, 180
squirrel, 81
star patterns, 31
starling, 58, 92, 183, 184
state space, 70, 71
stickleback, 23, 72, 73, 138, 140, 169
  ten-spined, 158
  three-spined, 4–6
stimulus, supernormal, 41
stimulus filtering, 47–50
sunfish, bluegill, 150, 151

taxes, 29
territory, 165–7
testosterone, 62, 92, 93, 103
Tets, G. F. van, 108, 109
thermoregulation, 114
Thorpe, W. H., 88
threat posture, 36, 38
tiger, 165
Tinbergen, Niko, 1, 2, 36, 38, 66, 67
tit
  blue, 184, 185
  great, 121, 124, 125, 184, 185
  long-tailed, 136, 137
toad
  calling, 152, 153
  prey capture, 45, 46

reproduction, 138, 139
vision, 85
treefrog, green, 142
Trivers, Robert, 139
typical intensity, 116

urine, 160

vacuum activity, 53, 58, 59
vision
  binocular, 43, 45
  brain pathways, 49
  communication, 158
  development, 84, 85
  human, 34

waggle dance, 161, 162
walking, 25
warbler
  garden, 28–31
  marsh, 92
  sedge, 154
warm up, 17, 18
Washoe, 163
wasp, 148
water boatman (*Notonecta*), 59–61
Watson, J. B., 76–9
weaning conflict, 98
weaverbird, red-billed, 181
weed dance, 117
whale, killer, 93
Whitman, Charles Otis, 1
wildebeest, 56, 172
Wilkinson, Jerry, 178
wolf, grey, 169
Wolda, H., 59–61
woodpecker, black, 80
Wynne-Edwards, V. C., 134, 135

Zanfurlin, Mario, 19, 20
zebra, plains, 172
zig-zag dance, 4, 5, 36, 38